MEAT AND FISH MANAGEMENT

Stephen A. Mutkoski
Cornell University,
School of Hotel Administration

Marcia L. Schurer
University of New Hampshire,
Department of Hotel Administration

Breton Publishers
A division of Wadsworth, Inc.
North Scituate, Massachusetts

Breton Publishers
2 Bound Brook Court
North Scituate, Massachusetts 02060

Library of Congress Cataloging in Publication Data

Mutkoski, Stephen A.
  Meat and fish management.

  Bibliography: p. 287
  Includes index.
  1. Food service management.   2. Meat.   3. Fish.
I.   Schurer, Marcia L.   II.   Title.
TX911.3.M27M87   642'.5'068   81-892
ISBN 0-534-00907-7   AACR2

Printed in the United States of America
  2   3   4   5   6   7   8   9   –   87   86   85   84   83

PREFACE

During my first few years teaching at the School of Hotel Administration, Cornell University, I searched for teaching aids, textbooks and films designed for use in a food service oriented meats course. While attending various seminars of the American Meat Institute, the American Meat Science Association and the National Restaurant Association I interviewed other educators to ascertain their needs and determine the educational material that they were currently using.

From the information gathered, it appeared that there are two completely different textbook approaches. The first is the animal science approach, which is extremely technical for most food service managers and students. The emphasis in this type of text is placed on breeding, and the chemistry and microbiology of meat. While texts of this nature are excellent reference works, they have no management orientation. There is little or no coverage of the basic areas of major concern to the food service manager, such as purchasing, receiving, preparation, and cost analysis.

The second approach is management-oriented. Texts of this sort, however, lack the technical information necessary to give management complete control. For example, the reader is instructed to use the USDA grades as a guide to quality in purchasing, but the specifications used in determining the grades is not provided. As a result, the reader is at a loss to evaluate an upgraded product or even to be sure of the quality level of the graded one.

There was clear evidence of the need for a single book to draw together both the managerial and technical aspects of meat, fish, and poultry. There are many food service oriented programs where there is demand for this approach. This prompted me to begin work on the book in conjunction with Ms. Marcia Schurer, a former graduate student of mine. Ms. Schurer has since lectured at several colleges and has held a full time position in the Department of Hotel Administration at the University of New Hampshire. Without her persistence this book might not be complete today.

We feel we have achieved an integration of the "meat science" approach and the "management" approach. In doing so, we hope this text will be useful both to students preparing for a career in food service management and to the food service managers, chefs, purchasing agents, etc., that are dealing day to day with this costly commodity.

STEPHEN A. MUTKOSKI, PH.D.
MARCIA L. SCHURER, M.P.S.

Photo Credits

CONTENTS

1 | MEAT SCIENCE AND MANAGEMENT, AN EDUCATIONAL APPROACH

The hospitality industry has long enjoyed the status of a frontier industry—there has been plenty of room for newcomers and revenues have continued to grow. The food service sector of the industry has shared in this munificence. But the last decade and a half has seen several economic factors colliding: the proliferation of competition, the rise in all costs, inflation, and recession with concomitant slowing, if not retrenchment of industry growth.[1] These factors have keened the competitive edge and made new demands on the professionalism and creativity of the industry.

The training of the decision makers and the workers is a critical part of this professionalism (and training at its best may contribute to this creativity too). There is no longer room for the trial-and-error on-the-job training of the past, for the dollars that those errors cost are too important now.

The element of training to which this text is addressed is the instruction of food service managers in meat science and management. The significance of this subject to the food service industry becomes evident when one examines the dollars spent on purchasing meat, fish, and poultry products. According to D. Sherf, it is common for a restaurant to spend from 35 to 45 percent of its food-buying dollars on meat, fish, and poultry.[2] In college feeding, the percentage is somewhat lower; D. Dwyer puts it between 28 and 32 percent.[3] T. Hazzard states that 40 percent of the purchasing dollar can be attributed to meat, fish, and poultry purchases.[4] Some steak house chains spend as much as 50 percent of their food-purchasing dollars for meat. (See Table 1.1.)

TABLE 1.1 PERCENTAGE OF FOOD-PURCHASING MONEY SPENT ON MEAT, FISH, AND POULTRY

Classification	Percentage
City hotels	40
College feeding	28–32
Commercial*	42
Motel*	38
Country clubs	45

* Hotels and motels classified "commercial" have more business people as guests than do those classified "motel." The latter are more family-oriented.

1

Depending on the nature of the business, the percentage spent on meat, fish, and poultry will vary, but in every case it will be the largest single contribution to food cost. Surely then, learning the technical aspects of meat, as well as the managerial aspects, is a crucial requirement in the training of anyone who seeks a career in food service management.

1.1 A BLEND OF TECHNOLOGY AND MANAGEMENT

In meat science and management technical knowledge is the critical input upon which food service management must be based. Identification of cuts of meat and knowledge of the criteria for grading are examples of technical knowledge. Application of this information toward the development of a set of purchasing specifications for a food service operation is an example of management.

Today, more than ever, both technical knowledge and management skills are needed. Today's food service manager does not normally have a full-time butcher on hand whose specialized skills in meat products and whose years of experience have given the knowledge necessary to handle the product properly. Managers must each bring the essential elements of this knowledge with them when they walk on the job because decisions on handling the product will be theirs to make.

The manager must be able to absorb the meaning of many technical changes that occur, such as changes in grading standards for beef. It is not sufficient for the food service manager to know about changes in terminology; one must understand what the changes mean in terms of the product. One must understand the compositional differences that have occurred as a result of the new grading standards—for example, that when **marbling** level is reduced, the fat is replaced by moisture. Then one must apply this factual knowledge to one's managerial policies in purchasing, storage, and preparation of the beef graded under the new system.

A manager without adequate training in meats is dependent on the goodwill of many people. Such a reliance can be very costly to the company or institution the manager represents. A purveyor can immediately sense a lack of expertise in weak purchasing and receiving practices, and take it as an invitation to unload inferior quality, short weights, and excess fat, all at premium prices. Employees within the food service operation can also be quick to take advantage of such an opportunity; purchasing agents and receiving personnel are in excellent positions to receive **kickbacks** from their purveyors or simply to pilfer meat. Without technical as well as managerial training, a manager would be at a loss to correct this situation; in fact, he or she might not even realize it exists. Food costs would continue to rise and quality to decrease until nothing would be left of the business.

1.2 PURPOSE AND OBJECTIVES

The purpose of this text is to supply the student and manager with the technical skills and information necessary to manage all phases of meat, poultry, and seafood in a food service operation. The text is written in a building block sequence with the technical areas presented first, followed by coverage of each major product—that is, beef, lamb, veal, and so on. This information is then applied to the management subsystems of a food service facility, including purchasing, receiving, storage, preparation, and cost analysis.

The major objective is to provide the manager with the knowledge and skills to:
1. Judge the **quality** of meat, poultry, and seafood
2. Maintain or enhance the quality through **proper storage**
3. Improve the product's **palatability** by employing proper tenderization methods
4. Identify **major wholesale** and **portion cuts** of meat, poultry, and seafood
5. Choose the most economical and marketable way to **utilize** each wholesale and portion cut
6. Select the most **appropriate cooking method** for each wholesale and portion cut
7. Conduct **fabrication** and **yield tests** on products
8. Perform a **cost analysis** of fabrication
9. Accurately make new **menu pricing** decisions
10. Write clear and concise product **specifications**
11. Develop adequate **policies and procedures** for purchasing, receiving, and storage of meat, poultry, and seafood products

1.3 STATUS OF THE MEAT INDUSTRY AND ITS EFFECT ON THE FOOD SERVICE INDUSTRY

Figure 1.1 (on the left) shows the historical flow of meat through various market channels to the food service industry. In early years, the **slaughterer** was the end of the distribution chain; hotels and restaurants would buy major **carcass cuts** (*sides, quarters*) direct from the slaughterer. This meant the purchaser had to have a butcher on the premises and a rather large selection of menu items to utilize all the by-products of fabrication.

As refrigeration and transportation improved, "breakers" came on the scene as the middlemen between the slaughterer and the food service industry. The **breaker** would purchase carcasses and quarters and "break them down" to (fabricate them as) major **primal** and **wholesale cuts** (ribs, rounds, loins, and so on). This allowed hotel and restaurant operators to deemphasize major fabrication and to reduce menu offerings somewhat.

Figure 1.1 Changes in meat industry supply channels.

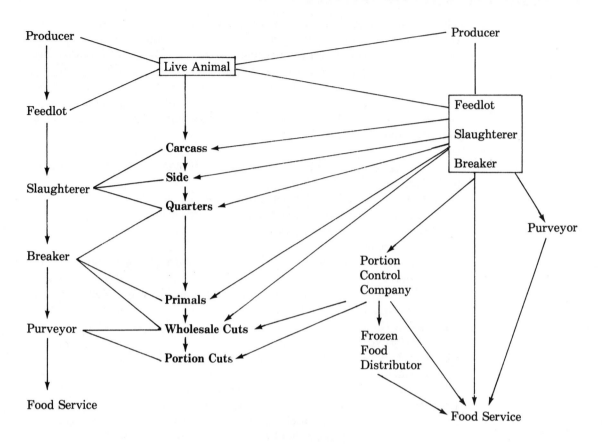

When **purveyors** entered the scene, they offered even more services. Gradually, they began to supply the major products needed by food service operations (beef, veal, pork, lamb, poultry). Purveyors also cut to order on specification, allowing the hotel and restaurant industry to specialize and allowing rib and steak houses, with their very limited menus, to appear.

With the advent of **boxed meats** in 1973, distribution channels were altered once again (see the right-hand side of Fig. 1.1. Large companies took over the functions of feedlot operator, slaughterer, and breaker all under one roof. Such operations feed the animal, slaughter it, and break it down to wholesale cuts that are vacuum packaged in a plastic film. The packaged wholesale cuts may then be shipped to a purveyor or directly to the food service industry. The economic advantages of this change are tremendous: first it eliminates shipping a high percentage of waste (fat, bone, hide); second, product loss due to shrinkage and deterioration is significantly reduced. Boxed meat has been readily accepted in the food service industry as a way to reduce cost and provide better shelf life.

Another supply segment of the meat industry, which actually began before boxed beef appeared but has assumed a much larger share of the market since then, is the **portion-control supplier**. This segment of the industry specializes in supplying such products as exact and uniform portion-cut steaks, chops, and patties. The acceptance of boxed meat has helped portion-control suppliers because, typically, they start with trimmed boxed meat and fabricate the portion cuts. As a result of modern freezing and packaging technology, one portion-control plant can ship its product to distributors throughout the world. Uniformity in quality, appearance, palatability, and cost are some of the advantages that portion control supply offers. As labor costs increase and the availability of skilled labor decreases, portion control cuts of meat will be utilized more and more by the food service industry.

This text was written with the recent marketing changes in mind. Although the carcass structure of each major type of meat is presented briefly at the start of each chapter, it is done so only to acquaint the reader with the location of the major wholesale cuts on the carcass. This information is necessary for the reader to understand why products have different structural components and why they differ in value and intended use. Emphasis is placed on the wholesale cuts and the portion cuts that are obtained from those cuts. This emphasis continues throughout the book.

REFERENCES

1. S.A. Mutkoski, "Guide to Curriculum Development Meat Science and Management Instruction for the Food Service Industry," *Ph.D. Thesis* (August 1976).
2. David Sherf, Former Partner and Consultant for Laventhol & Horwath. Executive Vice President, Hotel Equity Financing Corporation.
3. Dan Dwyer, Former Purchasing Agent for Cornell Dining.
4. Ted Hazzard, Purchasing Agent for the New York Hilton Corporation.

2 | COMPOSITION AND STRUCTURE OF MEAT AS A FOOD

The structure and composition of meat are generally not the favorite topics of students or food service managers. Nonetheless, information on these topics is a necessary basis for understanding and controlling important technical and managerial aspects of meat. For example,

- Proper selection of cuts for specific menu items is based on their muscle structure and amount of connective tissue. For example, the **tenderloin**, a long, fine-grained muscle with little **connective tissue**, is an expensive cut generally used for top quality steak called filet mignon. Using this cut to prepare a beef stew would be a waste of money. The **shoulder clod**, a coarser muscle containing a higher percentage of connective tissue, could not make a very tender steak but would produce an economical stew.
- Wholesale and portion cuts are identified through knowledge of structural characteristics such as bone, shape of muscle, and fat cover.
- Properly applying tenderization and cooking methods depends on the structure of the muscle.
- Spoilage can be avoided or minimized by the manager who is aware of the major chemical reactions that take place in the muscle and who recognizes the benefits or dangers of these reactions. (Enzymatic reactions can create a tenderer product, but if uncontrolled they can destroy the integrity of the product.) Storage and inventory policies must take these reactions into consideration.
- Judging quality of meat requires knowledge of the characteristics that indicate quality, like marbling, color of the lean, and texture of the muscle.
- To fabricate properly (that is, to cut up the carcass into roasts, steaks, chops, and so on), bone structure and muscle grain must be understood and the location of connective tissue sheaths (natural seams) between muscles must be recognized.

The technical information provided in this chapter will be applied throughout the remaining chapters.

Meat is generally considered to be the flesh of any animal used for food. Beef, lamb, veal, pork, poultry, fish, and game all are meats by this definition. The U.S. Department of Agriculture (USDA) further defines meat as the edible part of skeletal muscles, as well as what are commonly known as variety meats—the tongue, diaphragm, heart, esophagus, and glandular meats from the brain, liver, kidney, thymus, spleen, pancreas, and stomach linings.

TABLE 2.1. COMPOSITION OF MEAT

Components	Percentage
Proteins	16–22
myofibrillar: muscle	
sarcoplasmic: enzymes and pigments	
stromal: connective tissue	
Lipids: fats	1.5–42
Bone and cartilage	0–26
Carbohydrates	1
Moisture	65–80
Vitamins	Minute
Minerals	0.5

Of the components listed in Table 2.1, the lean muscle is the most important and makes up the bulk of what we call meat. The connective tissue acts as a structural component to hold the muscles together and to attach them to the bone. The bone and cartilage supply a framework and support for the entire carcass. Fat layers protect the internal organs, cushion and insulate the surface of the skeletal muscles, and are an important determinant of palatability when dispersed within the muscle. Moisture, carbohydrates, vitamins, and minerals are carried in the muscle and fat constituents of meat and will be discussed under those headings.

2.1 MUSCLES

Meat is composed mainly of muscle. Muscles vary in location, composition, size, shape, function, and the amount of connective tissue associated with them. Each muscle contains fibers of characteristic size related to the total size of the muscle and the amount of work required of it. The **grain** or texture of meat is determined by the size of the fibers that make up the muscle. Muscles engaged in gross movement, like the muscles of the chuck (a cut from the shoulder area) develop larger **fibers**, a coarser texture, and as a result are naturally less tender than muscles engaged in little activity, like the tenderloin. Because the tenderloin functions more as a support muscle and is not engaged in strenuous activity, it has small muscle fibers, a smooth texture, and is quite tender.

Types of Muscles

Three types of muscles are found in the animal body: **voluntary striated, involuntary striated,** and **involuntary smooth.** The voluntary striated, or skeletal muscles, represent the bulk of an animal's muscles and are the most widely used in food service. They represent the muscles of the rib eye, strip loin, tenderloin, and top round (see Fig. 5.1 in Chapter 5, Beef). The heart is an example of an involuntary striated muscle. Involuntary smooth muscle may be found in such areas as the blood vessels and viscera. Because the skeletal or striated muscle is the major type, it will be examined in some detail in the following sections.

TABLE 2.2   APPROXIMATE COMPOSITION OF MAMMALIAN SKELETAL MUSCLE (PERCENT FRESH WEIGHT BASIS)

	Percentage		Percentage
Water (range: 65 to 80 %)	75.0	Nonprotein Nitrogenous substances	1.5
Protein (range: 16 to 22%)			
Myofibrillar	9.5	Creatine and creatine	
Myosin	5.0	phosphate	0.5
Actin	2.0		
Tropomyosin	0.8	Nucleotides	
Troponin	0.8	(Adenosine triphosphate (ATP),	
M protein	0.4	adenosine diphosphate (ADP),	
C protein	0.2	etc.)	0.3
α-actinin	0.2	Free amino acids	0.3
β-actinin	0.1	Peptides	
Sarcoplasmic	6.0	(anserine, carnosine, etc.)	0.3
Soluble sarcoplasmic and		Other nonprotein substances	
mitochondrial enzymes	5.5	(creatinine, urea, inosine	
Myoglobin	0.3	monophosphate (IMP),	
Hemoglobin	0.1	nicotinamide adenine	
Cytochromes and flavo-		dinucleotide phosphate	
Proteins	0.1	(NADP)	0.1
Stroma	3.0	Carbohydrates and nonnitrogenous	
Collagen and recticulin	1.5	substances (range: 0.5 to 1.5%)	1.0
Elastin	0.1	Glycogen (variable range: 0.5	
Other insoluble proteins	1.4	to 1.3%)	0.8
Lipids (variable range: 1.5		Glucose	0.1
to 13.0%)	3.0	Intermediates and products of	
Neutral lipids (range: 0.5		cell metabolism	
to 1.5%)	1.0	(hexose and triose phosphates,	
Phospholipids	1.0	lactic acid, citric acid, fumaric	
Cerebrosides	0.5	acid, succinic acid, acetoacetic	
Cholesterol	0.5	acid, etc.)	0.1
		Inorganic constituents	1.0
		Potassium	0.3
		Total phosphorus	
		(phosphates and inorganic	
		phosphorus)	0.2
		Sulfur (including sulfate)	0.2
		Chlorine	0.1
		Sodium	0.1
		Others	
		(including magnesium, calcium,	
		iron, cobalt, copper, zinc,	
		nickel, manganese, etc.)	0.1

From *Principles of Meat Science* by John C. Forrest, Elton D. Aberle, Harold B. Hedrick, Max D. Judge, and Robert A. Merkel. W.H. Freeman and Company. Copyright © 1975.

Composition of Skeletal Muscles

Skeletal muscles are composed of moisture, protein, fat, carbohydrate, nitrogenous compounds, and inorganic constituents. Table 2.2 shows the percentages of each of these constituents.

CLASSIFICATION OF MUSCLE PROTEINS The most important component of muscle tissue is protein. Skeletal muscle consists of three basic types of proteins:

1. **Sarcoplasmic proteins** constitute the enzymes and pigments within the muscle.
2. **Stromal proteins** are comprised of collagen, elastin, and reticulin.
3. **Myofibrillar proteins** are the contractile and regulatory elements of the muscle.

2.2 MYOFIBRILLAR PROTEINS

The myofibrillar proteins make up the largest percentage of the muscle proteins. The structure of the muscle and how it functions can best be grasped by starting with the structure of the myofibrillar proteins. The sequential organization of the muscle is presented in chart form in Fig. 2.1.

Figure 2.1. Structure of muscle

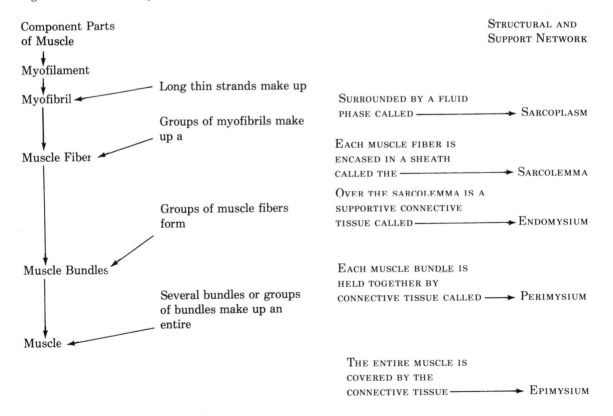

Groups of myofilaments form myofibrils. Groups of myofibrils encased in the sarcoplasm form muscle fibers, which are in turn surrounded by the sarcolemma. These muscle fibers vary in length and diameter and are supported by connective tissue, called endomysium. Groups of muscle fibers form bundles, or fasciculi, which are enclosed in perimysium. Groups of muscle bundles form entire muscles and are enclosed in connective tissue referred to as the epimysium. (See Figs. 2.2 and 2.3.)

Figure 2.2   Diagrammatic sketch of a muscle fiber. From Price and Schweigert, The Science of Meat and Meat Products, *W. H. Freeman and Company, San Francisco, 1960, p. 16.*

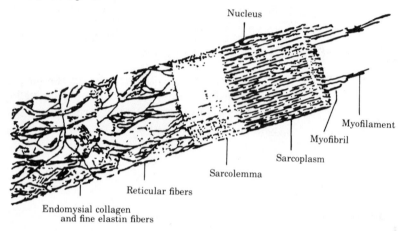

Figure 2.3   Muscle cross section illustrating the arrangement of connective tissue into epimysium, perimysium, and endomysium, and the relationship to muscle fibers and fasciculi. The typical position of blood vessels is also shown. From Price and Schweigert, The Science of Meat and Meat Products, *W. H. Freeman and Company, San Franscisco, 1960, p. 24.*

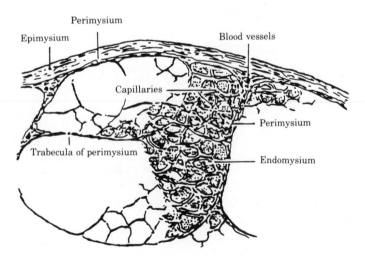

Striated Muscle

It is within the myofibrils that the patterns of light and dark bands commonly called muscle striations are found. When viewed under a microscope, the myofibrils are made up of rows of **sarcomeres**. (See Fig. 2.4.)

Figure 2.4 A drawing adapted from an electron microscope showing portions of two myofibrils and a sarcomere (×15,333) and a diagram corresponding to the sarcomere, identifying its various bands, zones, and lines. Modified from "The Mechanism of Muscular Contraction," by H. E. Huxley, Scientific American, December 1965.

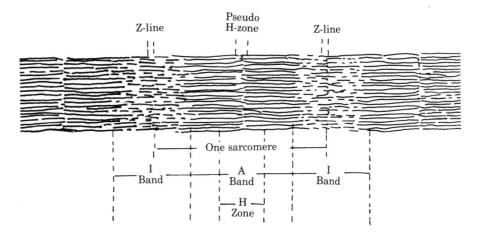

Each sarcomere consists of a **Z line** or disk, which holds the muscle together, **one-half an I band**, then an **A band** with an **H zone** in the middle, then another one-half I band and another Z line or disk. The A bands are typically dark, whereas the **I bands** are light. Located in the bands of the sarcomere are the myofibrillar proteins, which are responsible for muscle contraction. These proteins, shown in Fig. 2.5, are **myosin**, in thick filaments primarily in the A band, which play a major role in muscle contraction and relaxation; **actin**, the thin filament that interacts with myosin; α-**actinin**, which is located in the Z line and thin filaments and which influences the cross-linking of actin; β-**actinin**, which is possibly located in the thin filaments and which deters the interaction among actin strands; and **M-proteins**, which are located in the middle of the H zone and which hold the thick filaments in position in the A band.

Myosin accounts for 50–55% of the myofibrillar proteins, actin 15–20%, tropomyosin 5–8%, troponin 5–8%, α-actinin 2–3%, β-actinin 0.5–1%, and M-protein 3–5%. As noted before, myosin and actin are the myofibrillar proteins most responsible for muscle contraction and relaxation, which in turn affect the tenderness of meat. The remaining proteins are regulatory proteins that turn contractions on and off in the presence of adenosine triphosphate, known as ATP.

Figure 2.5   Sketch of sarcomere model indicating probable location of several proteins.

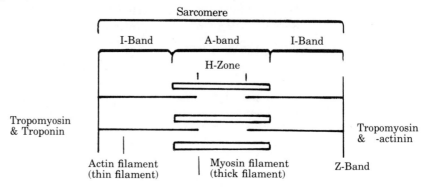

Muscle Contraction in Live Animals and Post Mortem

In the resting state actin and myosin are free to slide over each other. The **magnesium complex** of **adenosine triphosphate (MgATP)** prevents the myosin heads from combining with the actin filaments. During contraction, the actin filaments are pulled actively over the myosin filaments by the myosin heads drawing them toward the center of the A band. Stimulus from the nervous system signals the release of calcium (Ca^{++}), the ATP breaks down to adenosine diphosphate (ADP), causing the cross-bridges of the myosin (the myosin heads) to attach to the thin filaments of the actin and forming the complex known as **actomyosin**. When the nervous system ends the signal, the calcium is removed. When ATP is formed again, the actin and the myosin are dissociated and the muscle returns to the relaxed state.

When an animal is slaughtered, normal muscle functions continue for a while until glycogen is depleted and the energy compounds such as creatine phosphate and ATP are depleted. At this point, permanent cross-bridges are formed between actin and myosin, forming the complex actomyosin as in muscle contraction. This results in the stiffening of the muscle known as **rigor mortis**. More **cross-bridges** are formed during rigor than occur during muscle contraction in live animals (in vivo muscle contraction), which accounts for the very rigid state of the **postmortem muscle** in rigor. Onset of rigor is a function of carcass size and type as well as the pH and temperature of the muscle. Rigor may begin in poultry a few minutes after slaughter, but in beef the time required may be six to twelve hours.

Muscles in the state of rigor are less tender than prerigor or postrigor (resolved) muscles. **Resolution of rigor** occurs with partial disintegration of the Z-line structure and reduction of the number of cross-bridges. This allows the sacromere to lengthen and it reduces the rigidity of the muscle. It is believed that the class of enzymes called cathepsins are partially responsible for resolution of rigor.

2.3 STROMAL PROTEINS: CONNECTIVE TISSUE

Connective tissues are composed of stromal proteins. As can be seen from the diagrams on muscle structure, connective tissues play a very important role in the support of the muscle. They also provide support and structure for organs, blood vessels, and nerves.

The three major types of connective tissue are **collagen, elastin,** and **reticulin**. All three are similar in composition: they contain relatively few cells, which are embedded in a ground substance or matrix to which extracellular fibers of collagen, elastin, or reticulin are attached.

The amount of collagen, elastin, or reticulin present in a muscle depends upon the location and activity of that muscle. The more work the muscle is required to do, the more connective tissue will be present.

Collagen

The most important connective tissue is collagen because it is the principal structural component of tendons, ligaments, cartilage, and bone. It is a whole tissue, present in every muscle. It varies in amount from 3 to 30 percent of the total muscle protein (depending on the use of the muscle). Collagen is the major component of the connective tissue sheaths (endomysium, perimysium, and epimysium) that encase muscle fibers, muscle bundles, and entire muscles. These sheaths are valuable aids in meat fabrication because they permit separation of the major muscles from the carcass without incising the muscle itself. They are also helpful in identifying the various muscles of the carcass because they maintain the characteristic shape of the muscle.

Collagen presents a problem when it comes to tenderness because connective tissue is considerably less tender than the muscle itself. Cuts of meat that have a higher percentage of connective tissue (like cuts from the chuck) are less tender than cuts from sections of the carcass that have little connective tissue (cuts from the loin, for example).

EFFECTS OF COOKING. Upon heating, collagen first shrinks and then hydrolizes to form gelatin. Moisture is required for gelatin formation to take place, so that cuts with a high percentage of collagen often require moist heat cookery. The age of the animal determines how much collagen will **hydrolize**. As the animal ages, the number of cross-linkages formed in the collagen molecule increases. This makes the collagen less soluble so that a lower percentage of collagen changes to gelatin when heated, which partially explains why older meat is tougher.

Elastin

A very dense, fibrous yellow tissue, elastin, is a structural component of ligaments, arteries, organs, and muscles. The ligament called the "**back strap,**" which connects the bones along the back to the vertebrae from the neck through the ribs is a good example of elastin. It is a very rubbery substance, which will not hydrolize when heated, even in the presence of water. Fortunately, the amount of elastin contained in most muscles is small.

Reticulin

Fibers of reticulin form a fine network around the surface of the sarcolemma, organs, muscle cells, and blood and lymplistic capillaries. Because reticular fibers resemble collagen, some researchers believe it to be a form of immature collagen.

2.4 SARCOPLASMIC PROTEINS

Pigments are the major class of sarcoplasmic proteins of interest here. The color of meat is primarily due to the muscle **heme** pigment (iron pigment), **myoglobin**, whose main function is to store oxygen in the animal's tissues. The blood heme pigment, hemoglobin, is found in the capillaries and is responsible for transporting the oxygen from the lungs.

The amount of myoglobin present in lean meat varies according to several factors. The species, breed, feed, sex, and age of the animal have a great effect on the color of the lean. As the animal grows older the amount of myoglobin present increases. This change helps to explain why veal (generally under three months of age) has a lean of light pink color compared to the bright red color of beef. Pork and veal contain 1–3 mg of myoglobin per gram of wet tissue, beef has 4–10 mg per gram, and older beef has 16–20 mg per gram. The amount may also vary depending on the type of muscle, activity level involved, and the amount of blood and oxygen available. Certain muscles of the same animal may therefore be darker in color than other muscles.

Color Changes

Myoglobin is greatly affected by exposure to air, heat, bacteria, enzymes, alcohols, and acids. All of these factors produce changes in the color of the lean, some desirable and some undesirable. When beef is cut, the surface of the lean is purple in color. The muscle heme pigment is in the reduced state (see Fig. 2.6 for the chemical reaction of myoglobin). As the lean is exposed to air, the myoglobin in the tissues absorbs the oxygen, becoming **oxymyoglobin** and changing the lean to a bright red color. This

Figure 2.6 Chemical reaction of myoglobin.

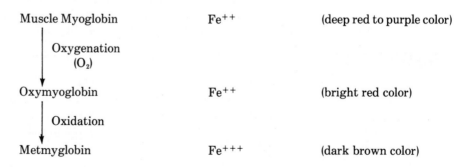

Muscle Myoglobin	Fe^{++}	(deep red to purple color)
Oxygenation (O_2)		
Oxymyoglobin	Fe^{++}	(bright red color)
Oxidation		
Metmyoglobin	Fe^{+++}	(dark brown color)

reaction is sometimes referred to as the bloom of fresh meat. When the meat is exposed for too long, however, or when it is exposed to bacteria, metals, heat, or freezing temperatures, the myoglobin heme is oxidized to an undesirable dark brown color as it becomes **metmyoglobin**.

WARNING SIGNS. If sanitation is poor and high numbers of bacteria are present, other color changes can occur. **Leuconostoc viridans** will produce an iridescent sheen on the surface of the product. Other **pseudomonads** will produce an undesirable reaction, causing the myoglobin to turn green. In some cured products, this greening reaction progresses to a loss of color state, where the product appears translucent. Although many of these changes are produced by bacteria that are **nonpathogenic** (they do not cause food poisoning), the bacteria may produce off-odors and off-flavors in the product. Color changes are important to the food service operator because they give an indication of (1) original quality of the meat when purchased and (2) sanitation and temperature control during processing and because they provide a warning sign to the operator if (3) the product is not being stored adequately or inventory is not being used fast enough.

2.5 LIPIDS: FATS

Meat contains two classes of lipids: (1) storage fats or fat depots and (2) intracellular muscle or structural lipids. **Triglycerides** predominate in both although some monoglycerides and diglycerides can be present. **Phospholipids** are found in small amounts particularly in the membranes. The types of fats and fatty acids deposited in the fat depots vary with the species of animal and location in the animal. Fatty acids present in the carcass can be grouped according to the saturated (**palmitic, stearic**) and unsaturated (**oleic, linoleic,** and **arachidonic**) states. Oleic (42 percent fatty acid in beef fat), palmitic (29 percent) and stearic (20 percent) occur in substantial amounts in the fat depots. Linoleic (2 percent) and arachidonic (0.1 percent) occur in much smaller amounts. Only 0.5 percent linolenic is present in beef.

Fat develops in and around the cells and between and over the muscles. Three stages of development are evident. Fat begins to develop first around the internal organs (the kidney, for example) as a protective cover. Later it spreads over the surface of the muscles to cushion and insulate, and finally it develops within the muscle. This last stage of fat development, intramuscular marbling, is the one that contributes the most to improving the palatability of the product.

High-quality cuts of meat should have a fine, even distribution of marbling throughout the muscle. This intramuscular fat contributes greatly to the aroma, flavor, and juiciness of the product. Whereas moderate amounts of marbling are desirable to ensure high levels of palatability, however, excess interior fat and surface fat are costly and should be avoided. (See the discussion of yield grading in Chapter 3.)

Meat fats vary in keeping quality and sometimes result in loss during storage. Two mechanisms are responsible for fat deterioration. The first is **hydrolysis** (the splitting of the fatty acid from glycerol), usually caused by enzymatic activity. The free fatty acids alone can produce off-flavors and off-odors.

The second is **oxidative rancidity**. Fatty acids are attacked by oxygen, producing off-odors and flavors. Unsaturated fatty acids, particularly those containing several double bonds, are more apt to degenerate rapidly. Proper packaging, temperature control, and inventory rotation can prevent losses due to oxidative rancidity. (See Chapter 16, on storage.)

2.6 BONES AND CARTILAGE

The connective tissue cells produce blood cells, blood-forming tissues, and cartilage and bone, which are also connective tissues. The bone structure is helpful in identification of cuts and in determining the age and the quality of the animal. In young animals, the bones are soft, porous, and bloody with signs of cartilage still evident in the area of the feather bones (also called buttons). The rib bones are round and the vertebrae are not fused. As the animal matures, the rib bones flatten out, becoming brittle and white in color, and the vertebrae fuse in the order sacral, lumbar, thoracic.

2.7 CARBOHYDRATES

Of the carbohydrates present in meat, the **glucose polymer glycogen** is the one of most interest to the food service operator. Although a small amount of glycogen is found in most animal tissues, the largest percentage is stored in the liver. Glycogen is a storehouse of energy for the muscle, and the amount present in the muscle at the time of slaughter affects the ultimate pH (acidity) of the muscle. A change in pH can alter the color, water-binding capacity, tenderness, and storage life of the muscle.

Importance of Glycogen Levels

When the animal is slaughtered normal respiratory functions continue for a short time until the remaining oxygen is depleted and the energy compounds of adenosine triphosphate (ATP) and creatine phosphate are used up. At this point energy is supplied anaerobically by the conversion of glycogen to lactic acid, with ATP being produced. This process is called glycolysis, and it accounts for the post mortem accumulation of lactic acid in the muscle. The amount of glycogen present in the muscle at the time of slaughter directly influences the quantity of lactic acid that will accumulate after the animal is killed. When living, the animal muscle has a pH of 7.5. After slaughter the pH drops to 5.3–5.9 as the lactic acid builds up.

DARK CUTTERS If the level of glycogen is low at the time of slaughter, however, little lactic acid will accumulate and the pH will be higher (6.0–6.6). The lower acidity causes the lean to darken considerably;

the resulting beef is called a "dark cutter." Normally, when beef is cut and exposed to air, the color changes from purple (myoglobin) to bright red (oxymyoglobin). This change does not occur when the surface of a dark cutter is exposed to the air. In that case the amount of oxygen absorbed is decreased and the myoglobin is only partially oxygenated.[1] Studies show that the lower the glycogen level, the lower the acidity, and the lower the oxidation-reduction potential, which produces beef of a darker shade of red.

The level of glycogen present at the time of slaughter depends on the degree and duration of energy demands. Studies have shown that the combination of inadequate feeding and shipping procedures prior to slaughtering lowers the glycogen content of the muscle, thus increasing the occurrence of dark cutter beef. Additionally, if the animal is exhausted or stressed at the time of slaughter, glycogen levels may drop, causing a darker lean.

Stress-susceptible pigs may become anaerobic more rapidly post mortem than normal pigs and as a result have a fast rate of glycolysis.[2] The rapid buildup of lactic acid to lower than normal pH levels (pH 5) decreases the water-binding capacity of the meat. This loss of water-binding capacity is exhibited in PSE (pale soft exudative) pork (see Chapter 8, Pork), which suffers unusual shrink losses upon cooking.

Ultimate pH effects on meat Esthetically the dark cutter is considered inferior to the normal lighter, bright red colored beef. There are other reasons for the food service operator to reject dark cutters. Beef with a high ultimate pH may be a better medium for bacterial growth,[3] thus reducing storage life. Researchers report that higher than normal pH values adversely affect tenderness; a decrease in tenderness is exhibited in meat as the pH increases from normal (5.5) to 6.0.[4] As the ultimate pH increases above 6.0, tenderness begins to increase again; however, at 6.8 and above the meat exhibits a gelatinlike texture that is not desirable.[5]

Lower than normal pH, as exhibited in PSE pork, is also undesirable because it adversely affects color and the water-binding capacity of the muscle. The lower ultimate pH produces a very pale light color and decreases the ability of the muscle to retain moisture, causing excessive shrink losses during cooking.

2.8 MOISTURE

Water is the major component of meat. It typically constitutes 60–70 percent of the muscle. Water affects the quality of meat, particularly the juiciness, tenderness, taste, and yield of cooked meat.

In view of the fact that water is the biggest constituent of meat, in percentage of weight, the food service operator must control the moisture loss in storage and cooking. If a 10-pound strip loin is purchased for $38.00, an operation cannot afford to lose 10 percent of the weight of that strip loin because of moisture loss. Storage and preparation policies and procedures must take moisture into consideration.

2.9 MINERALS AND VITAMINS

Minerals help to build the skeletal system, bones, and teeth and act as normal constituents of the soft tissues and fluids to keep the body properly regulated. Vitamins are necessary for life to promote growth, protect health, and to prevent specific deficiencies.

The major minerals present in meat are **calcium, phosphorus,** and **iron.** About 99 percent of the calcium is located in the bones. Calcium is an activator of enzymes and is involved in the contractile process of muscles. Phosphorus is located in the bones (80 percent), blood cells, and acids involved in metabolism of carbohydrates.

There are two classes of vitamins: **fat-soluble** and **water-soluble.** Of the fat-soluble vitamins (A, D, E, K), vitamin A and D are found in larger quantities in the liver than elsewhere. Of the water-soluble vitamins, the B complex, especially niacin and riboflavin are supplied by meat. Pork provides much more thiamine than other meats. From a nutritional standpoint, meat has a high biological value because it supplies all of the essential amino acids, which are the building blocks of protein and which are required to sustain life.

2.10 SUMMARY

This chapter has provided the fundamental scientific knowledge a food service manager needs to understand in order to control meat, fish, and poultry in a commercial food service system. The technical terms and concepts are the basis on which information in the following chapters is built.

Practical applications of the technical information on composition and structure will be given in the discussions of grading, palatability, purchasing, and other topics. Referring back to this chapter as specific examples are discussed should give the reader a better understanding of the examples and an appreciation of the benefits of a knowledge of composition and structure of meat for the food service manager.

CASE STUDY

A shipment of product has just been delivered to your restaurant facility by Sam's Meat Packing House. The shipment consists of various kinds of meat—beef, veal, pork, lamb—with several different cuts of each kind.

During a routine inspection of the product at the receiving dock, the receiving clerk has observed the following: There is a big difference in the color of each product. Even within the same kind of meat, some cuts are much lighter in color than other cuts and a few are extremely dark in color. One beef cut has a greenish tinge to it.

The receiving clerk has also noticed a distinct difference in the coarseness of the muscle grain in one particular carton that contained all the same type of muscle (beef chucks). In comparing these chucks to other cuts he found the chucks had more sheaths of white tissue separating the muscle into groups.

In a carton of beef ribs he found the rib bones were round and contained a high amount of blood. Sections of the bone were still soft.

Using the information provided in this chapter, discuss the receiving clerk's findings, explaining what structural and chemical components account for the characteristics observed. Consider the implications of these characteristics. What effect will they have on the value of the products to the restaurant?

1. What causes the color differences between the different species?
2. What causes color difference within a species, for example, veal versus beef?
3. What causes color difference within a single carcass?
4. Why does a given muscle change color?
5. What do these color changes tell us about product quality and condition?
6. What accounts for different size muscle fiber and grain coarseness?
7. How does this affect product quality?
8. Why did the chuck show more sheaths separating the muscles than other cuts?
9. How does this affect product use and quality?
10. What is the significance of the condition of the rib bones?

REFERENCES

1. C.A. Winkler, *Canad. J. Res. D.* (1939) 17:8.
2. E.J. Briskey, "Recent Points of View on the Condition and Meat Quality of Pigs for Slaughter," p. 41 (eds. W. Sybesma, P.G. Van Der Wals and P. Walstron), *Res. Inst. Animal Husbandry,* (1969).
L.L. Kastenschmidt, *The Physiology and Biochemistry of Muscle as a Food,* (eds. E.J. Briskey, R.G. Cassens and B.B. Marsh), (Madison: Univ. Wisconsin Press, 1970), vol. II: 735.
3. M. Ingram, *Ann. Inst. Pasteur,* (1948) 75:139.
4. P.E. Bouton, F.D. Carrol, P.V. Harris and W.R. Shorthose, *J. Fd. Sci.* (1973) 38:401.
5. P.E. Bouton, A. Howard and R.A. Lawrie, *Spec. Rept. Fd. Invest. Bd.,* (1975) No. 66.

3 | GOVERNMENT REGULATIONS: INSPECTION AND GRADING

The food service industry is in the business of serving the public. Public health and safety are of great importance when it comes to the products served in a hotel, a restaurant, or an institutional food facility.

The potential hazards of transmitting disease, causing an outbreak of food poisoning, or accidentally poisoning people with products containing toxic chemicals should be of concern to all food service operators.

Government regulations requiring inspection of meat products at various stages of processing, good manufacturing practices, and the monitoring of products for hazardous residues provide the food service industry with an important assurance of product safety.

Besides being concerned with product safety, however, food service managers must also be concerned with the quality and yield of the products purchased. Consistency in product quality, in such attributes as tenderness, juiciness, and flavor, as well as consistency of product yield, which is the number of servable portions remaining after trim, are essential to maintain customer satisfaction and profitability. Government grading programs provide the food service industry with quality and yield ratings on meat products. They are very useful guidelines for purchasing meat.

In this chapter federal inspection programs will be discussed first because inspection is a prerequisite for federal grading. United States Department of Agriculture (USDA) **quality** and **yield** grades will be explained and **packer grading** systems, house grades of meat packing companies, will also be investigated.

3.1 OBJECTIVES OF GOVERNMENT INSPECTION

The most devastating thing that could happen in a food service operation would be an outbreak of food poisoning or disease caused by the products sold in that facility. It is for this reason that the federal inspection program is an extremely valuable safeguard to the hospitality industry and the consumer.

The **Federal Meat Inspection Act** and the **Poultry Product Inspection** Act, which date from 1957 as well as the more recent **Wholesome Meat Act**, passed in 1967, were all designed primarily to protect the consumer. The inspection program reaches its objectives by ensuring the consumer of wholesomeness and cleanliness as well as accuracy of measures and labels of meat and meat products.

Wholesomeness

All meat, meat products, and processed items entering the food distribution chain must be free from disease. Hence animals slaughtered and processed for consumption are monitored by government inspectors to ensure that they are free from disease and free as well from harmful ingredients such as pesticides, herbicides, antibiotics, and any other dangerous chemical residues that might be present in their bodies.

Sanitation

Adequate sanitation is necessary in the slaughtering and processing of meat products for consumption; therefore, many aspects of the processing are inspected to ensure that the bacterial level of the product is minimized. The personnel working in a processing plant must be clean and appropriately attired to perform their various jobs. The plant itself is constantly inspected. The condition of equipment used in the plant is checked, and standards of cleanliness are established for each area. The procedures used in the processing of meat for consumption are continuously checked by federal inspectors. Temperatures within the plant and its storage facilities are monitored to guarantee that they are cold enough to retard bacterial growth. And the product as it passes through the slaughtering and processing operations is constantly checked to make sure that it is handled in a sanitary manner.

Accuracy of Measures and Labels

Federal inspection gives the consumer some assurance that weights and measures are correct. Requiring accuracy of labeling of products prevents false or misleading statements from being printed on the label. It is also a safeguard against the use of additives that might falsify the actual condition of the product.

The examples in the following list illustrate the USDA standards established for certain meat and poultry products sold in commerce. Standards for some products are more rigid than for others and may specify processing or cooking procedures as well as the minimum percentage required for a product to be labeled under the names given.[1]

MEAT PRODUCTS (All percentages of meat are on the basis of fresh uncooked weight unless stated otherwise.)

Beef Sausage (Raw): No more than 30% fat. No by-products, no extenders, and no more than 3% water.

Breaded Steaks, Chops, etc.: Breading must not exceed 30% of finished product weight.

Frankfurter, Bologna, and Similar Cooked Sausage: May contain only skeletal meat (see Chapter 2). No more than 30% fat, 10% added water, and 2% corn syrup. No more than 15% poultry meat (exclusive of water in formula).

Ham—Cooked or Cooked and Smoked (Not Canned):   Must not weigh more after processing than the fresh ham weighs before curing and smoking; if the item contains up to 10% added weight, it must be labeled "Ham, Water Added"; if it contains more than 10% added weight, it must be labeled "Imitation Ham."

Hamburger, Hamburg, Ground Beef, or Chopped Beef: No more than 30% fat; no extenders.

Scallopine: At least 35% meat (cooked basis).

Veal Steaks: Can be chopped, shaped, or cubed. Beef can be added with product name shown as "Veal Steaks, Beef Added, Chopped, Shaped, and Cubed" if no more than 20% beef, or must be labeled "Veal and Beef Steak, Chopped, Shaped, and Cubed." No more than 30% fat.

POULTRY PRODUCTS   (All percentages of poultry (chicken, turkey) are on a cooked deboned basis unless stated otherwise.)

Canned Boned Poultry, Boned (Kind), with Broth: At least 80% poultry, meat skin, and fat.

Chicken Cordon Bleu: At least 60% boneless chicken breast (raw basis), 5% ham, and either Swiss, Gruyere, or Mozzarella cheese. (If breaded, no more than 30% breading.)

The Inspectors

The **USDA Food Safety and Quality Service** is the governmental agency responsible for the administration of the federal meat inspection program. The national health inspection service employs approximately 9,000 veterinarians and other inspectors assigned to some 6,800 meat and poultry slaughtering and processing plants throughout the country. Inspectors who are not licensed veterinarians must complete a training program and be certified for the position.

3.2 INSPECTION PROCEDURES

There are three distinct phases of the inspection procedure: 1) *ante mortem* (before death) inspection of the animals; 2) *post mortem* (after death); 3) inspection throughout fabrication and processing.

Ante Mortem

Prior to entering the slaughtering facility the live animal is examined by the federal inspector for any visible signs of disease. The inspector judges the overall condition of the animal, looking for signs of illness, shipping

TABLE 3.1 ANTE MORTEM INSPECTION

Species	Passed	Suspected	Condemned	Total
Cattle	36,621,976	117,670	14,550	36,814,196
Calves	3,470,302	1,289	19,045	3,490,636
Sheep	7,543,599	2,859	9,025	7,555,483
Goats	62,576	5	95	62,676
Swine	70,922,008	11,420	73,358	71,006,786
Equine	260,598	65	343	261,006
Total	118,881,059	193,308	116,416	119,190,783

Source: *Statistical Summary, Federal Meat and Poultry Inspection for Fiscal Year 1976.* Meat and Poultry Inspection Program, Animal and Plant Health Inspection Service, U.S. Department of Agriculture (January 1977).

fever, unusual growth, malformation, irregular discharges, and other problems. One of three things may happen at this point. If the animal is free of any visual sign of illness, the inspector passes the animal and allows it to be slaughtered. If it has some localized infection, the animal is tagged "suspect" and kept separate from the normal, healthy animals. Any "suspect" animal is then examined very closely to judge the nature of the localized infection. If the animal is not judged healthy enough to be released for slaughter, the third and most drastic action is taken: the inspector then condemns the animal. This means that the animal would not be allowed to enter the slaughtering plant. The carcass from a condemned animal would have to be disposed of without coming in contact with meat products that have been passed, that is, approved for consumption. In some cases, after the animal is destroyed, the entire carcass must be buried or burned. Table 3.1 shows how many animals were passed, suspected, and condemned in the United States during the fiscal year of 1976 during the ante mortem inspection.

Post Mortem

The second major procedure in inspection occurs post mortem. After the slaughter, the federal inspector examines the carcass very thoroughly. The head, lungs, heart, liver, spleen, and all other internal organs as well as the lymphatic system are checked for any visible signs of disease. Once again one of three decisions may be made. The carcass may be passed for consumption, it may be retained as "suspect" and each major division of the carcass inspected separately, or the entire carcass may be condemned, requiring its removal from the carcasses that passed inspection. Table 3.2 shows how many animal carcasses were passed and how many were condemned during post mortem inspection in the United States for the fiscal year 1976.

TABLE 3.2 POST MORTEM INSPECTION

Species	Passed	Condemned	Total
Cattle	36,688,607	110,120	36,798,727
Calves	3,452,415	18,790	3,471,205
Sheep	7,506,653	39,764	7,546,417
Goats	62,390	188	62,578
Swine	70,813,553	119,237	70,932,790
Equine	259,546	1,110	260,656
Total	118,783,164	289,209	119,072,373

Source: *Statistical Summary, Federal Meat and Poultry Inspection for Fiscal Year 1976.* Meat and Poultry Inspection Program, Animal and Plant Health Inspection Service, U.S. Department of Agriculture (January 1977).

Inspection throughout the Fabrication and Processing

The inspection job is not complete at the point of slaughter but continues throughout the entire processing to ensure that the product is handled in a sanitary fashion, that weights and labels are accurate, and that any ingredients added to the product are wholesome and do not falsify the product's appearance or condition.

The cost of federal inspection is paid from taxes. The rate paid a federal inspector is approximately $19 per hour. The cost per pound of meat inspected is estimated to be approximately 0.44 cents, or less than ½ cent per pound. The federal inspectors are paid by the U.S. Department of Agriculture, not by the packer or processor.

The following statistics show the importance of the federal meat and poultry inspection program. In 1975, federal meat inspectors took the following actions[2]:

- Inspected 3 billion birds, 113 million meat animals, 69 billion pounds of processed products
- Condemned 37 million birds, 390,000 meat animals, 60 million pounds of processed products
- Reviewed over 127,600 product labels and rejected almost 818,000 labels
- Tested 19,600 samples for residues
- Ran almost 87,500 tests to identify species of meat and to determine what bacteria, antibiotics, and other substances might be present
- Passed for U.S. entry into the United States more than 1.7 billion pounds of foreign products and rejected about 12.5 million pounds
- Detained over 11.3 million pounds of "suspect" meat and poultry in 931 detention actions

3.3   INSPECTION IS MANDATORY

Any meat product sold for commercial use must be inspected. In some areas of the country, state inspection plants are still in existence. These plants must conform to the standards of the federal inspection program. Plants operating under a state's inspection can sell the product only within that state. Federal inspection is required on all meat shipped interstate, any meat sold under government contract, and all imported and exported meat products. In addition, more than 1,000 plants in other countries are certified by the USDA to ship products to the United States. The foreign plants are monitored by U.S. inspectors at frequent intervals to guarantee consistent compliance with our regulations. The product is further inspected at the port of entry prior to being distributed in the United States.

Products processed under federal inspection bear one of the three stamps illustrated in Figure 3.1 to signify that they have been inspected and passed. The federal inspection stamp assures the buyer that the meat product is wholesome, that it has been processed in a sanitary manner, and that any statements on the label concerning weights or ingredients are accurate. The federal inspection stamp gives no indication of the quality of the product purchased, however; the assigned grade does that.

Figure 3.1 USDA inspection stamps.

In the mark that appears on primal and major wholesale cuts the number identifies the packaging plant and the abbreviation stands for "Inspected and Passed."

The mark that appears on prepackaged processed meat products, sold mostly retail, is also found on processed items going to wholesale distribution.

Poultry products that have been federally inspected bear the same mark whether fresh, frozen, and processed.

3.4 USDA GRADING SYSTEMS: QUALITY AND YIELD

The food purchasing agent for a hotel or restaurant appreciates the guarantees of wholesomeness provided by the federal inspection program. But there are other aspects of meat purchasing that are also very important for a buyer. The purchasing agent must be capable of selecting a product that will meet the needs of his market. Quality and price are the buyer's primary considerations in placing an order.

The USDA **quality grading** system gives the buyer a standard by which to purchase meat. This system attempts to rate the overall palatability of a given product, thereby providing the buyer with a preview of how good a particular meat product will taste to the consumer. It is a rating system that predicts the flavor, juiciness, and tenderness of the product prior to consumption.

The USDA **yield grading** system provides the buyer with a standard for judging the economy of a particular meat product. The yield grade is assigned in an attempt to predict the amount of usable meat that will be obtained in proportion to the amount of waste that must be trimmed from a given carcass or wholesale cut. These ratings on palatability and yield are extremely important indicators for the purchasing agent. Table 3.3 shows the USDA quality grades and yield grades.

Need for Thorough Knowledge of Grading Characteristics

The food service manager or purchasing agent must know more than just the terminology used for the quality grading system. That is, it is not enough to know only the names of the grades; one must know what the grades represent and be able to assess quality and yield oneself. The individual responsible for buying meat products must know what goes into grading a product. The receiving clerk must be trained to recognize the characteristics that should be present in a given grade or quality. This knowledge is particularly important today for a number of reasons.

- Often when **portion-controlled steaks** and chops are trimmed to comply with the purchaser's specifications, the USDA grade stamp is cut off. The receiving clerk must be knowledgeable enough to determine the actual grade level of the product.

TABLE 3.3   USDA QUALITY AND YIELD GRADES

Beef	Yield Grades for Beef and Lamb	Lamb	Veal and Calf	Pork
Prime	1	Prime	Prime	U.S. #1
Choice	2	Choice	Choice	U.S. #2
Good	3	Good	Good	U.S. #3
Standard	4	Utility	Standard	U.S. #4
Commercial	5	Cull	Utility	U.S. Utility
Utility			Cull	
Cutter				
Canner				

- Many large packers are promoting their own **house grades**. Should the buyer decide to purchase these products, the buyer's representative must be knowledgeable enough to rate the products and make comparisons with products bearing the USDA grade.
- Some carcasses never receive a grade of any type. Purchasing agents selecting from ungraded carcasses must be aware of the characteristics that indicate quality.

Without an accurate assessment of quality, the product may be put to the wrong use, resulting in the customer's dissatisfaction. A USDA Choice strip loin will produce juicy, flavorful, and tender steaks for broiling, but a strip loin of USDA Commercial grade would not possess the same palatability characteristics.

3.5 QUALITY GRADING OF BEEF

In the case of beef, the carcass is chilled after inspection and then cut between the twelfth and thirteenth rib to expose the rib eye muscle. This muscle is used by the USDA grader to rate the entire beef carcass. The characteristics used as criteria for judging the quality of the carcass are as follows: (See color insert on next page.)

1. Texture of the lean
2. Color of the lean
3. Firmness of the lean
4. Marbling or fat dispersed within the lean
5. The age of the carcass

TEXTURE OF THE LEAN The rib eye muscle from the top grades—Prime and Choice—has a very fine-grained texture, smooth to the touch. Texture is an indication of how much the muscle has been used. As the animal increases in age and exercises more, the muscles generally become coarse in texture. This coarse texture makes the cooked beef less tender.

COLOR OF THE LEAN The color of the lean gives an indication of the age of the animal. The meat from young bovine, called veal, has a very light pinkish-colored lean. In high-quality beef, from animals ranging in age from 16–24 months, a bright cherry red color is desirable. In older, more mature cattle, the color of the lean continues to darken to a much deeper red because of increased amounts of myoglobin within the muscle tissue resulting from more use of the muscle. The color changes, therefore, give an indication of the age of the animal and make color an extremely important criterion in rating quality. Again, generally speaking, the older the animal, the less tender its muscles will be because of the extended period of use. Figure 3.4 illustrates how the color of the lean becomes progressively darker with maturity and the texture becomes coarser.

Figure 3.2 Identifying the quality grades of beef.

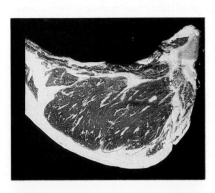

PRIME
Characteristics
Color of the lean: bright cherry red
Texture of the lean: fine
Firmness of the lean: firm
Rib bones: slightly wide and flat
Marbling: slightly abundant to abundant
Chine bone: slightly red and soft
Cartilage on end of thoracic vertebrae: some ossification of
  lumbar vertebrae (loin section)
Sacral vertebrae: completely fused

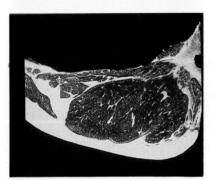

CHOICE
Characteristics
Color of the lean: bright cherry red to moderate red
Texture of the lean: fine
Firmness of the lean: slightly firm
Rib bones: slightly wide and flat
Marbling: small to moderate to slightly abundant
Chine bone: slightly red to tinged with red
Cartilage on end of thoracic vertebrae: partially ossified
Sacral vertebrae: completely fused
Note: The major difference between Prime and Choice is the
amount of marbling.

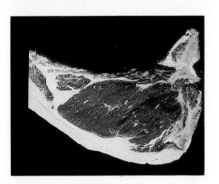

GOOD
Characteristics
Color of the lean: slightly cherry red to slightly dark red
Texture of the lean: fine
Firmness of the lean: moderately to slightly soft
Rib bones: slightly wide and flat
Marbling: slight to small amount
Chine bone: slightly red to tinged with red
Cartilage on end of thoracic vertebrae: partially ossified
Sacral vertebrae: completely fused

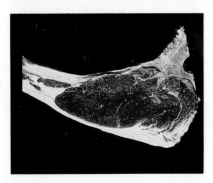

STANDARD
Characteristics
Color of the lean: light to slightly dark red
Texture of the lean: fine to moderately fine
Firmness of the lean: soft to moderately soft
Rib bones: slightly wide and flat
Marbling: practically devoid of marbling to traces of it to
  slight amount
Chine bone: slightly red to tinged with red
Cartilage on end of thoracic vertebrae: partially ossified
Sacral vertebrae: completely fused

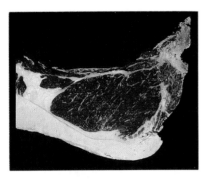

COMMERCIAL
Characteristics
Color of the lean: moderately to dark red
Texture of the lean: slightly coarse to coarse
Firmness of the lean: slightly firm to firm
Rib bones: moderately wide and flat to wide and flat
Marbling: ranges from small to modest to moderate to
  slightly abundant
Chine bone: moderately hard to hard and white
Cartilage on end of thoracic vertebrae: barely visible
Note: On first glance, the amount of marbling might lead one
to rate this rib high. But the texture, color of the lean, and
the yellowish fat as well as the bone condition identify it as
Commercial to the experienced eye.

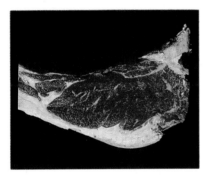

UTILITY
Characteristics
Color of the lean: slightly dark red to very dark red
Texture of the lean: fine to coarse
Firmness of the lean: soft and slightly watery to slightly firm
Rib bones: slightly wide and flat to wide and flat
Marbling: ranges from practically none to traces to slight to
  small to modest to moderate amount
Chine bone: slightly soft and red to white and hard
Cartilage on end of thoracic vertebrae: some ossification to
  barely visible
Sacral vertebrae: completely fused

CUTTER
Characteristics
Color of the lean: slightly dark red to very dark red
Texture of the lean: fine to coarse
Firmness of the lean: soft and watery to soft and slightly watery
Rib bones: slightly wide and flat to wide and flat
Marbling: ranges from practically none to traces to slight amount
Chine bone: slight red and soft to hard and white
Cartilage on end of thoracic vertebrae: completely ossified to barely
  visible
Sacral vertebrae: completely fused

CANNER
*Only those carcasses that have less than the minimum requirements
necessary for the Cutter grade are graded Canner.*

Color changes during cooking indicate degree of doneness, as shown in Figure 3.7. (See discussion in Chapter 17, p. 250.) Figure 3.8 shows that evenness in degree of doneness results from low-temperature cooking of roast beef.

FIRMNESS OF THE LEAN   Higher quality is indicated by a firm but elastic lean muscle. In lower quality beef the muscle is soft to mushy to the touch. The amount of firmness gives an indication of the type of feed used when the animal was raised. Grass-fed cattle usually have a softer, less firm lean than grain-fed cattle.

MARBLING   This term refers to the particles of fat dispersed within the lean of the muscle. A great deal of emphasis was placed on marbling in the past, when it was believed that increased marbling could offset the toughening generally associated with advanced maturity in beef. However, marbling has only a slight effect on tenderness, especially in young beef animals. It is more important for its effect on flavor and juiciness. In higher grades of beef, an abundance of marbling, finely dispersed throughout the entire muscle, is desirable. The photos in Figure 3.3 illustrate the levels of marbling used in grading beef.

AGE OR MATURITY   The age of the animal is an important aspect of grading. The USDA grader determines the age by checking the bone structure of the carcass. In a young beef animal (16–22 months of age), the following characteristics might be expected: red porous rib bones, fairly round in shape, with evidence of soft cartilage apparent at the end of the featherbones ("buttons"). As the animal increases in age, the redness dissipates from the ribs and the rib bones tend to flatten out and become white in color. The cartilage in the buttons becomes firmly developed bone.

The spinal column can also be examined to judge the age of the animal. In general, **bone ossification** starts with the **sacral vertebrae** (sirloin area) and, as the animal ages, proceeds toward the front of the animal into the **lumbar vertebrae** (short loin area) and finally into the **thoracic vertebrae** (rib area). The USDA grader uses the bone structure in conjunction with the color of the lean and texture of the lean to predict the age of the animal.

See the photographs in Figure 3.2 illustrating the various grades of beef and the lists of their characteristics.

Figure 3.3 Marbling. Illustrations of the lower limits of certain degrees of typical marbling referred to in the official United States standards for grades of carcass beef (set by the United States Department of Agriculture, Agricultural Marketing Service, Livestock Division). Illustrations adapted from negatives furnished by New York State College of Agriculture, Cornell University. Degrees: 1) very abundant; 2) abundant; 3) moderately abundant; 4) slightly abundant; 5) moderate; 6) modest; 7) small; 8) slight; 9) traces; practically devoid not shown).

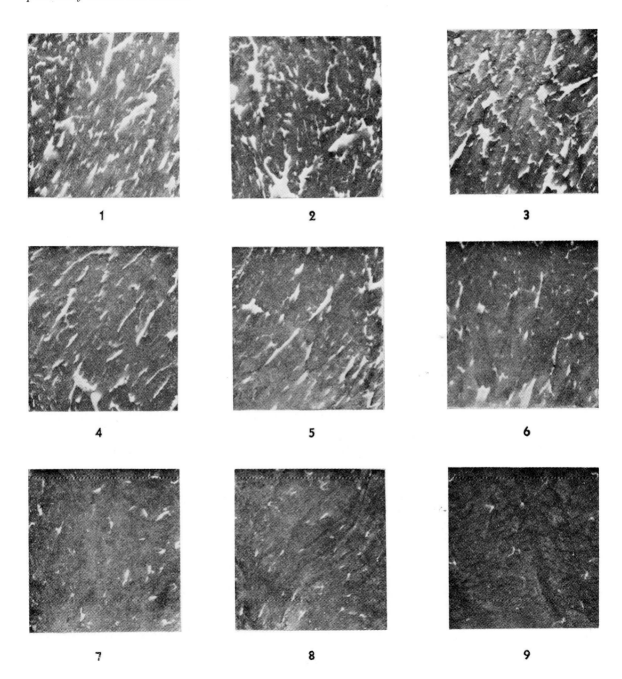

Figure 3.4 Darkening and coarsening of meat with age.

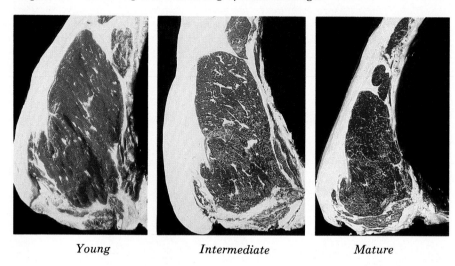

Young　　　　　　Intermediate　　　　　　Mature

Color also indicates the species of the animal. The photos in Figure 3.5 depict the typical color of lamb, veal, pork, and beef.

Figure 3.5 Color as an indication of species.

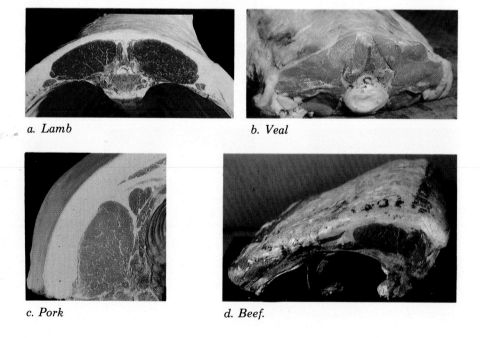

a. Lamb　　　　　　　　　　　*b. Veal*

c. Pork　　　　　　*d. Beef.*

Color reactions in fresh and processed meats, shown in Figures 3.6 a–e, can give the food service operator an indication of the condition of the product. (See discussion of myoglobin, Chapter 2, page 14, and of cured products, Chapter 13.)

Figure 3.6   Color reactions in fresh and processed meats.

a.   (left to right) Purple color of myoglobin; bright red color of oxymyoglobin; and brown color of metmyoglobin.

b.   Spare ribs, showing undesirable color.

c.   (left to right) Fresh pork, cured pork.

d.   Beef rib beginning to show signs of surface deterioration.

e.   Beef rib showing advanced signs of surface deterioration.

3.6 BEEF GRADING CHANGES

In February 1976, the USDA made three major changes in the beef grading system. The first change was to remove from the quality grading system the characteristic called **conformation**, which describes the shape and muscle development of the carcass. Because this characteristic has no bearing at all on palatability, it was removed from the grading system. It remains an important aspect in yield grading the carcass, however.

The second major change in beef grading was to require that all carcasses receiving a quality grade designation also be given a yield grade. Formerly, yield grading was completely optional; it was left up to the purveyor to decide whether or not he wanted the service of yield grading. Now in order for a product to be quality graded Prime, Choice, Good, and so on, it must also be yield graded. The yield grade stamp appears on the same ribbon roll as the quality grade stamp. The USDA instituted this change in order to encourage both consumer and packer to become more aware of and concerned about animals that had been fattened too much, resulting in carcasses with too much waste. It was hoped that this awareness would put more pressure on the feed lot operators to produce more efficient cattle—that is, more cattle that would grade Prime and Choice but with less trimmable fat.

The third aspect of the grading system to be altered was the marbling requirements for the youngest classification of beef-type animals listed as being of "A" maturity, or under 30 months of age. In the past, there was a heavy emphasis on the marbling requirement for the top grades of beef, for it was believed that extra marbling could offset the decrease in tenderness that often results from advanced age of the animal. Recent studies in animal husbandry have shown that in the meat from very young cattle, a certain minimal level of marbling is required to produce a satisfactorily tender product. Therefore, the USDA has taken a straight line approach to marbling with this "A" maturity grouping. Any animal reaching "B," "C," or any other stage of maturity must still show an increase of marbling to stay within a given grade. These changes are illustrated graphically in Figures 3.11 and 3.12, two charts of the relation between marbling, maturity, and quality as it was prior to the grading change and after the change.

Before the change (see Figure 3.11), a very young beef animal (16 months of age) needed only a slight to small amount of marbling to be graded Choice. That same animal at 30 months of age would require an increase in marbling up to the modest range in order to be graded Choice. If the marbling level remained at the slight to small range, the 30-month old carcass would then drop down in grade from Choice to Good.

As shown in Figure 3.12, any carcass under 30 months of age now requires only the minimal amount of marbling (slight to small) to be graded Choice. Therefore, a 30-month-old carcass that was graded Good under the old grading requirements would be graded Choice under this system. Note that these charts also illustrate the variation within a grade. Top Choice has a moderate amount of marbling whereas Low Choice has only a small amount.

In effect, the changes that took place in the beef grading system put the top portion of Good into the Choice range and the top portion of Choice into the Prime range. This increased the Prime category by approximately 5.8%. It also increased Choice by 8.9% and decreased Good by 6.2%. It has been estimated that these changes have reduced the marbling level in Prime and Choice from 12–14% to 8–10% on the average. This reduction of marbling may or may not affect tenderness within this young maturity range, but once the amount of fat within the muscle is reduced, only moisture will replace it. Therefore, this grading change may have a substantial effect on the flavor and juiciness of the products within the marginal Choice range. This factor may prove to be the one of most concern to the food service operator.

3.7 GRADING IS OPTIONAL

Slaughterers and packers are under no obligation to have their product federally graded. If a packing company wishes to have its product federally graded it must pay for the services of a USDA grader. The company may choose to have all of its products federally graded or instruct the grader to grade only those carcasses that would fall into the Prime and Choice range. Then the purveyor may sell the remainder of its products either ungraded or with some type of house grade.

In 1979, of all the cattle federally graded, 5.9% were Prime, 89.5% were Choice, and 3.8% were Good. When beef is graded it is almost always graded on the basis of the entire carcass. Occasionally individual wholesale cuts might be graded separately; it is also possible to have cuts from the same carcass graded differently; but such grading is the exception rather than the rule.

If the carcass has been graded by a federal grader with the kidney and pelvic fat intact, the ribbon roll of stamped grade marks would run from the chuck toward the primal round. If the carcass is graded after the kidney and pelvic fat have been removed, the ribbon roll would be placed on the carcass in the opposite direction, running from the neck toward the hindshank (see Chapter 5, Beef for anatomy). See Figure 3.13.

Packer Grades

Many large meat packers now merchandise their own house grades. The packers benefit by this system since they do not have to pay a fee for a USDA grader. They may also group their carcasses into whatever grade classification they see fit. However, they are not allowed to use any of the terminology used by the USDA in grading. Table 3.6 is a sample of brand names used by some of the major packers throughout the country.

Figure 3.7   Degrees of doneness.

a.   *Very rare*

b.   *Rare*

c.   *Medium rare*

d.   *Medium*

e.   *Well done*

Figure 3.8   Evenness in degree of doneness resulting from low-temperature cooking of roast beef.

a.

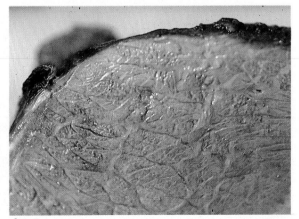

b.

3.8   YIELD GRADING OF BEEF

Four factors are used to predict the yield of a carcass: 1) the carcass weight; 2) the percentage of kidney, pelvic, and heart fat; 3) the area of the rib eye muscle; and 4) the thickness of the fat over the rib eye muscle. The yield grades range from 1 through 5 within each quality grade. A yield grade of 1 is ideal because it yields the most usable meat in relation to the amount of waste that will probably be trimmed off. A carcass of yield grade 5 is an extremely wasteful carcass and may contain 20 percent more waste than a #1 yield carcass of the same weight.

Figure 3.9 illustrates the importance of yield grades for those who purchase beef carcasses or primal and wholesale cuts. The center photograph shows the yield grade mark which is included in the same stamp as the quality grade mark.

It is easy to visually compare the yield grade 2 and yield grade 4 ribs shown in Figure 3.9. The statistics shown below give another indication of the amount of usable meat available from each. The yield grade 2 rib, from approximately the same carcass weight, yields 2.6 square inches more usable muscle meat than the yield grade 4 rib. The additional fat cover over the yield 4 rib is simply trimmable waste and does not increase the palatability of this rib at all. According to Tables 3.4 and 3.5, which compare yield grades, the difference in retail value per hundredweight between a yield grade 2 and a yield grade 4 carcass is $11.56; the yield grade 2 is a better buy because more usable meat remains after the fat is trimmed off.

Comparing yield grade 2 and yield grade 4 ribs shown in Figure 3.9

Yield Grade 2 (2.5)
Carcass weight 650 lb
Adjusted fat thickness 0.4 in.
Area of ribeye 12.7 sq in.
Estimated % kidney, pelvic, and heart fat 3.0%

Yield Grade 4 (4.5)
Carcass weight 620 lb
Adjusted fat thickness 0.8 in.
Area of ribeye 10.1 sq in.
Estimated % kidney, pelvic, and heart fat 4.0%

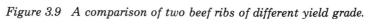

Figure 3.9   A comparison of two beef ribs of different yield grade.

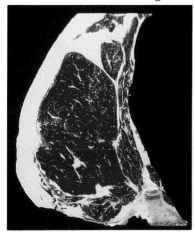

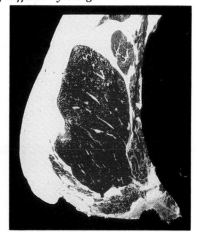

Rib from Yield Grade 2 carcass

Rib from Yield Grade 4 carcass

Importance of Yield Grade

The economic importance of yield grades to the food service operator becomes obvious when Tables 3.4 and 3.5 are studied. If a food buyer specifies simply USDA when ordering ribs of beef, that buyer is leaving himself or herself wide open to be shipped yield grade 4 or 5. Although the packer or purveyor has met the individual's specifications by supplying a USDA Choice quality grade, the buyer's food service operation is going to have a higher portion cost because of money spent on excess, or nonusable waste.

Many food buyers are not familiar with the yield grades. Anyone purchasing carcass, primal, and major wholesale cuts should specify which yield grade is desired. Portion-cut items such as steaks and chops do not generally require a yield grade because their purchasing specifications should include detailed requirements concerning fat trim, length of tail, thickness of the muscle, and other aspects. This type of detail would supersede the requirement for listing a yield grade.

To summarize what a government grader looks for in evaluating a carcass for both quality and yield, a USDA beef carcass evaluation report form is represented in Figure 3.10 for your information and reference.

Figure 3.10   USDA beef carcass evaluation report form.

USDA NO.	OTHER IDENTIFICATION	BREED *(As supplied by owner)*	MEAT GRADING CERTIFICATE NO.
NAME OF PRODUCER		NAME OF PACKER	

1 **QUALITY GRADE** ____ BY THIRDS	**A. CONFORMATION, MARBLING, AND MATURITY FACTORS**		
	CONFORMATION	DEGREE OF MARBLING	MATURITY (APPROXIMATE AGE SHOWN) *(Circle one)* **A B C D E** *(Under 30 mos.) (30 to 48 mos.) (Over 48 mos.)*

B. OTHER FACTORS

TEXTURE OF MARBLING *(Check one)*

☐ FINE   ☐ MEDIUM   ☐ COARSE

COLOR OF LEAN *(Check one)*

☐ VERY LIGHT CHERRY RED   ☐ CHERRY RED   ☐ SLIGHTLY DARK RED   ☐ MODERATELY DARK RED   ☐ DARK RED   ☐ VERY DARK RED   ☐ BLACK

FIRMNESS OF LEAN *(Check one)*

☐ VERY FIRM   ☐ FIRM   ☐ MODERATELY FIRM   ☐ SLIGHTLY SOFT   ☐ SOFT   ☐ VERY SOFT   ☐ EXTREMELY SOFT

TEXTURE OF LEAN *(Check one)*

☐ VERY FINE   ☐ FINE   ☐ MODERATELY FINE   ☐ SLIGHTLY FINE   ☐ SLIGHTLY COARSE   ☐ COARSE   ☐ VERY COARSE

2 **YIELD GRADE** ____ BY TENTHS	**YIELD FACTORS**			
	CARCASS WEIGHT *(From packer's hot wt. tag)* ____ LB.	FAT THICKNESS *(Inches, nearest 1/10 in.)* ____ IN. ACTUAL ____ IN. ADJUSTED	RIB EYE AREA *(from Grid)* ____ SQ. IN. BY TENTHS	KIDNEY, PELVIC AND HEART FAT *(As percent of carcass weight)* ____ PCT. ESTIMATED

____ (DATE)   ____ (SIGNATURE OF GRADER)

Figure 3.11 Pre-1976 chart of the relation between marbling, maturity, and quality (prior to grading change).

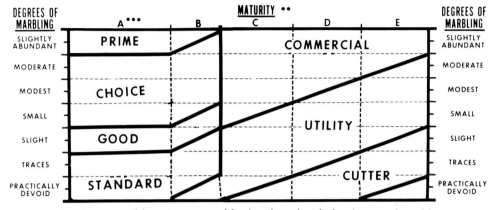

RELATIONSHIP BETWEEN MARBLING, MATURITY, AND CARCASS QUALITY GRADE ·

*Assumes that firmness of lean is comparably developed with the degree of marbling and that the carcass is not a "dark cutter."
**Maturity increases from left to right (A through E).
***The A maturity portion of the Figure is the only portion applicable to bullock carcasses.

Figure 3.12 Post-1976 chart of the relation between marbling, maturity, and quality (after the grading change).

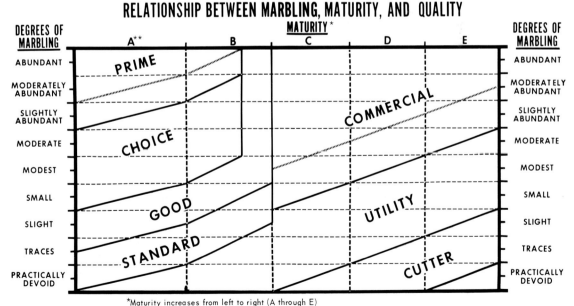

RELATIONSHIP BETWEEN MARBLING, MATURITY, AND QUALITY

*Maturity increases from left to right (A through E)
**The A maturity portion of the Figure is the only portion applicable to bullock carcasses.
........Represents midpoint of Prime and Commercial grades.

TABLE 3.4 YIELD GRADE COMPARISONS

If a retail operation sells 30,000 pounds of retail cuts and it bought carcasses of these USDA yield grades:	USDA Yield Grade #1	USDA Yield Grade #2	USDA Yield Grade #3	USDA Yield Grade #4	USDA Yield Grade #5
Then it would need to buy this much carcass beef:	36,585 lb	38,760 lb	41,210 lb	43,990 lb	47,170 lb
If, for example, the operation bought 600-lb carcasses it would need this many:	61	65	69	73	78
After preparing and trimming the retail cuts it would have the following yields of					
Fat trim	7.5%	12.6%	17.7%	22.8%	27.9%
Bone and shrink	10.5%	10.0%	9.5%	9.0%	8.5%
Trimmed cuts	82.0%	77.4%	72.8%	68.2%	63.6%
At current retail prices, the retail sales value per hundredweight (100 lb) of carcass would be:	$189.66	$180.06	$170.45	$160.84	$151.24
If all the carcasses were USDA Choice and were bought at the same price—$70.00 per hundredweight—the gross margin of profit as a percentage of sales would be*:	40.91%	37.83%	34.40%	30.58%	26.28%

Source: Marketing Bulletin # 45, USDA Yield Grades for Beef, U.S. Department of Agriculture.
*The cutting and trimming costs increase as the carcasses increase in fat. If these costs are included, the differences in profit margins and sales values between yield grades will be even greater than shown.

TABLE 3.5 COMPARISON OF YIELDS OF RETAIL CUTS AND RETAIL SALES VALUES CHOICE BEEF CARCASSES, BY YIELD GRADE*

Yield Grades		1		2		3		4		5	
Retail Cut	Price per pound	% of carcass	Value/ Cwt. carcass	% of carcass	Value/ Cwt. carcass	% of carcass	Value/ Cwt. carcass	% of carcass	Value/ Cwt. carcass	% of carcass	Value/ Cwt. carcass
Rump boneless	2.65	3.7%	9.80	3.5%	9.28	3.3%	8.75	3.1%	8.22	2.9%	7.68
Inside round	3.07	4.9	15.04	4.5	13.82	4.1	12.59	3.7	11.36	3.3	10.13
Outside round	2.82	4.8	13.54	4.6	12.97	4.4	12.41	4.2	11.84	4.0	11.28
Round tip	2.82	2.7	7.61	2.6	7.33	2.5	7.05	2.4	6.77	2.3	6.49
Sirloin, boneless	3.46	4.8	16.61	4.5	15.57	4.2	14.53	3.9	13.49	3.6	12.46
Short loin	3.82	5.3	20.25	5.2	19.86	5.1	19.48	5.0	19.10	4.9	18.72
Blade chuck	1.59	9.9	15.74	9.4	14.95	8.9	14.15	8.4	13.36	7.9	12.56
Rib, short cut (7")	2.73	6.3	17.20	6.2	16.93	6.1	16.66	6.0	16.38	5.9	16.11
Chuck, arm boneless	2.15	6.4	13.76	6.1	13.12	5.8	12.47	5.5	11.82	5.2	11.18
Brisket, boneless	2.43	2.5	6.08	2.3	5.59	2.1	5.10	1.9	4.62	1.7	4.13
Flank steak	3.69	.5	1.84	.5	1.84	.5	1.84	.5	1.84	.5	1.84
Lean trim	1.96	14.9	29.20	13.8	27.05	12.7	24.89	11.6	22.74	10.5	20.58
Ground beef	1.62	13.3	21.55	12.2	19.76	11.1	17.98	10.0	16.20	8.9	14.42
Kidney	.68	.3	.20	.3	.20	.3	.20	.3	.20	.3	.20
Fat	.11	8.1	.89	13.3	1.46	18.5	2.04	23.7	2.61	28.9	3.18
Bone	.03	11.6	.35	11.0	.33	10.4	.31	9.8	.29	9.2	.28
Total		100.0	189.66	100.0	180.06	100.0	170.45	100.0	160.84	100.0	151.24

Difference in retail value between yield grades -- $9.60 per cwt. of carcass.

Source: *Market News: Weekly Summary and Statistics*, Agricultural Marketing Service, U.S. Department of Agriculture, Washington, D.C.
*Difference in retail value between yield grades is $9.60 per hundred pounds of carcass. As yield grade goes from 1 to 5, the yield of each trimmed retail cut is less, making it less valuable. The difference in retail value between a yield 1 carcass and yield 5 is ($189.66–$151.24) $38.42 per hundred pounds.

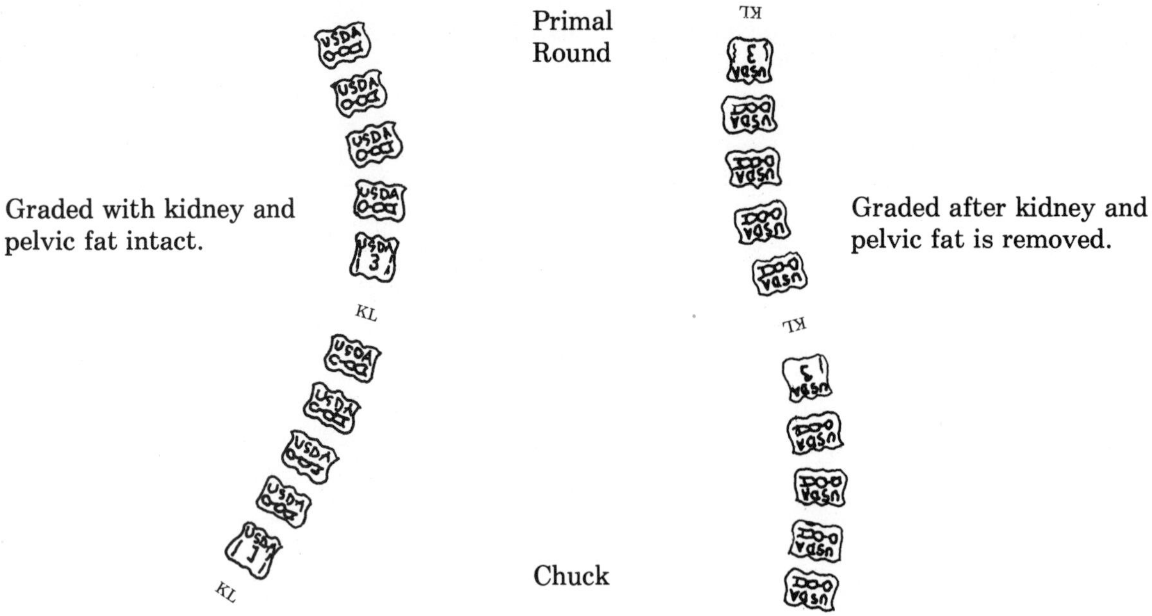

Primal
Round

Graded with kidney and
pelvic fat intact.

Graded after kidney and
pelvic fat is removed.

Chuck

*Figure 3.13 USDA inspection
stamp and grade.*

Figure 3.14 Hotel rack of lamb.

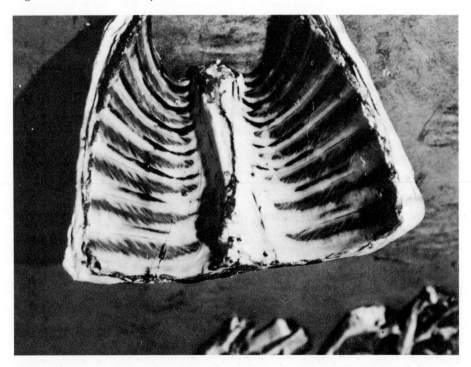

TABLE 3.6 IDENTIFIABLE BRAND NAMES USED BY VARIOUS MEAT PACKING COMPANIES FOR DRESSED BEEF

Packing Company	Brand Names*
Armour	Star Deluxe, Star, Quality, Banquet
Cudahy Foods (Division of General Host)	Puritan, Fancy, Rex, Rival, Thrifty-oh
George A. Hormel	Hormel Best, Hormel Merit, Hormel Value, Hormel
Hygrade Food Products	Peerless, Favorite, Economy, Holsum, Fair Reliable, Circle-K, Kimberly
John Morrell	Morrell Pride, Morrell Famous, Morrell Special, Morrell Allrite
Rath Packing	Black Hawk Deluxe, Black Hawk, Kornland, Crest, Sycamore†
Tobin Packing	Tobin Deluxe Choice, Tobin Blue Ribbon Choice, Tobin Selected
Swift	Swift Premium Proten, Swift Tru Tendr, Swift Premium, Swift Select
Wilson	Certified, Special, Ideal, Leader, Wilsco

* All brand names apply to steer and heifer beef. Listed in order of decreasing desirability.
† Crest and Sycamore are grades for cows.

The food service operator who purchases products with a packer's house grades can generally save a few cents per pound on these items. But one must take into consideration the integrity of the packing company and be able to judge the consistency of quality of the product when comparing it to the USDA graded alternative. It is extremely important that purchasing agents and receiving clerks who work with packer graded products have a good fundamental knowledge of the characteristics used to rate the quality of meat products.

3.9 LAMB GRADING

Lamb is both quality graded and yield graded. Many of the characteristics discussed earlier, in the section on beef quality grading, apply to the lamb carcass. The major difference between the two is that the lamb carcass is not "ribbed down" (cut between the twelfth and thirteenth rib) prior to grading. The USDA grader must therefore look at the interior of the carcass to make a quality determination. The following factors are judged by the grader: 1) the amount of **feathering**, that is, fat that has developed between the rib bones; 2) the amount of **cavity fat**, that is, the amount of fat that has developed over the chine bone, and 3) the presence of marbling, or streaks of fat, within the flank area of the carcass. The presence of fat in each of these three areas tells the USDA grader that the lamb carcass will have a substantial amount of marbling within the lean. In evaluation of ribbed lamb carcasses or wholesale cuts of lamb, the same factors that were examined for grading beef would be judged. Figure 3.14 shows the cavity fat and the feathering on either side of the rib bones of a hotel rack of lamb.

Determining the age of lamb is done by inspecting the bone structure of both the front legs and the rib bones (see Figure 3.15). A young lamb has thin, round, bloody rib bones, and a **break joint** at the end of the front legs. Mature lamb, also called mutton, has wider rib bones, the bones have less blood, and there is a smooth **spool joint** at the end of the front legs.

Figure 3.15 Break joint (left) of young lamb, and spool joint (right) of mature lamb, or mutton.

Break Joint   Spool Joint

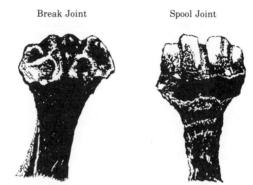

Lamb is yield graded for the same reasons as beef. The USDA grader looks at three characteristics to determine the yield grade:

1. The external fat over the twelfth and thirteenth ribs
2. The percentage of pelvic and kidney fat
3. The leg conformation score for the carcass (blockiness and development of muscles)

Table 3.7 illustrates the importance of yield grades in lamb.

TABLE 3.7   LAMB CHART COMPARING YIELDS*

CHOICE LAMB

Retail Cut	Price per pound	1 % of carcass	1 Value/ Cwt.	2 % of carcass	2 Value/ Cwt.	3 % of carcass	3 Value/ Cwt.	4 % of carcass	4 Value/ Cwt.	5 % of carcass	5 Value/ Cwt.
Leg, short cut	2.49	23.6%	58.76	22.2%	55.28	20.8%	51.79	19.4%	48.31	18.0%	44.82
Sirloin	3.17	6.7	21.24	6.4	20.28	6.1	19.33	5.8	18.39	5.5	17.44
Short loin	4.15	10.4	43.16	10.1	41.92	9.8	40.67	9.5	39.42	9.2	38.18
Rack	3.84	8.1	31.10	7.9	30.34	7.7	29.56	7.5	28.80	7.3	28.03
Shoulder	2.34	24.9	58.26	23.8	55.69	22.7	53.12	21.6	50.54	20.5	47.97
Neck	1.38	2.2	3.04	2.1	2.90	2.0	2.76	1.9	2.62	1.8	2.48
Breast	1.22	9.8	11.96	9.8	11.96	9.8	11.96	9.8	11.96	9.8	11.96
Foreshank	1.84	3.5	6.44	3.4	6.25	3.3	6.07	3.2	5.89	3.1	5.70
Flank	1.79	2.3	4.12	2.3	4.12	2.3	4.12	2.3	4.12	2.3	4.12
Kidney	1.24	0.5	.62	0.5	.62	0.5	.62	0.5	.62	0.5	.62
Fat	.11	4.6	.51	8.2	.90	11.8	1.30	15.4	1.69	19.0	2.09
Bone	.03	3.4	.10	3.3	.10	3.3	.10	3.2	.09	3.0	.09
Total		100.0	239.31	100.0	230.36	100.0	221.40	100.0	212.45	100.0	203.50

Difference in retail value between yield grades -- $8.95   per cwt. of carcass.
*The comparisons reflect average yields of retail cuts from beef and lamb carcasses typical of the midpoint of each of the USDA yield grades and average prices (including salepriced items) for USDA Choice beef and lamb during SEPTEMBER, 1980   .

Source: *Market News: Weekly Summary and Statistics*, Agricultural Marketing Service, U.S. Department of Agriculture, Washington, D.C.

3.10 PORK GRADING

The grading system for pork (USDA #1, 2, 3, and 4) combines the quality grading system with the yield grading system. All pork graded #1 to #4 has acceptable palatability characteristics. The major difference between the grades is the amount of usable meat obtained from each of them. However, a carcass that does not have quality characteristics that are acceptable would be graded U.S. Utility.

Four trimmed wholesale cuts are used to determine the grade designation. A USDA #1 pork carcass yields more than 53% of the carcass weight in the cuts of the fresh ham, the loin, the picnic, and the Boston butt. A USDA #2 yields between 50% and 53% in these four cuts, a USDA #3 yields between 47% and 50% of the carcass weight, and a USDA #4 yields less than 47% of the carcass weight in these four cuts.

The grades for pork carcasses are probably the least important to the food service operator since very few operations purchase whole pork carcasses. Cuts purchased wholesale are basically the same as the major retail cuts of pork. Therefore, trim designations take care of any excess fat that might occur on a U.S. #3 or #4.

3.11 VEAL AND CALF GRADING

The quality grades for veal and calf are the same: Prime, Choice, Good, Standard, Utility, and Cull. The main difference between the two is their maturity. **Veal** is from **bovine** animals under 3 months of age. **Calf** is bovine from 3 months to 7 months of age, at which point the classification becomes baby beef. The color of the lean and the bone structure are the most reliable characteristics used to differentiate between veal and calf.

Like lamb, veal is generally graded with the carcass remaining whole. The important characteristics used in grading are color, maturity, conformation, and the amount of feathering. Figure 3.16 shows the relationship between maturity and feathering.

Today there is one aspect of purchasing veal that may be more important than the USDA grades. It is the type of veal purchased (see Chapter 6, Veal).

The USDA Grader

The USDA grader is an employee of the Food Safety and Quality Division of the Department of Agriculture. The wages of the grader are paid by the USDA, but the purveyor is billed for these services. The grader is an individual who has been trained to rate the carcass after a visual inspection of several characteristics.

Figure 3.16   Relation between age and amount of fat in veal and calf.

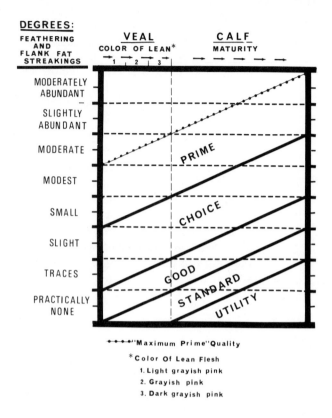

3.12   SUMMARY

From the illustrations shown in this chapter and the discussions of the procedures used to rate the quality and yield grade characteristics, one can readily see that there is a wide range of product quality and economy in the marketplace.

The USDA quality and yield grades are useful guidelines for purchasing, but they are not perfect. Two factors make this true: 1) Grading is a subjective evaluation by an individual, and human error is possible. 2) Cattle are raised, not manufactured, and each carcass differs in its makeup. Because of this variability, a range of acceptability exists within each grade. For example, beef carcasses with small, moderate, or modest amounts of marbling can be found within the Choice grade range. The food service manager should keep this range of acceptability in mind in order to select the level of the grade appropriate for his or her operation.

CASE STUDIES
1. You are the manager of a resort hotel and your purchasing agent has just come to you with an idea that she feels could save the operation money. Her brother-in-law raises cattle and sheep and is willing to slaughter the animals on his farm and deliver the dressed carcasses to the resort at a price below wholesale cost.

   Discuss the implications of such a purchasing agreement on product safety and quality.

2. SAM's packing house offers the purchasing agent of the firm you work for a price break on ribs of beef if he will purchase the "house" or "packer" grade. SAM assures the purchasing agent that the ribs are equivalent to USDA Choice and that the five cents per pound saved is additional profit. SAM states that all his ribs come from custom-fed cattle and that they are all delicious. The purchasing agent decides to try this house-graded product and notifies you, the receiving clerk to check it thoroughly to determine if it is equivalent to USDA Choice.

   As the receiving clerk, what do you look for to rate these ribs? What other factors besides quality should be considered in this type of purchase?

REFERENCES

1. *Standards for Meat Poultry Products: A Consumer Reference List*. Meat and Poultry Inspection Program, Food Safety and Quality Service, USDA (July 1977).
2. *The Federal Meat and Poultry Inspection Program: A Matter of Wholesomeness*. Animal and Plant Health Inspection Service, U.S. Department of Agriculture (December 1976).

4 | THE PALATABILITY OF MEAT

The food service manager's ultimate concern is the guests' satisfaction. Eating is a sensual experience, and the food service manager must be aware of the factors that affect this experience—in particular, palatability. One must be able to select products that have the right degree of palatability for one's market. One must also be able to maintain and in some cases enhance the overall palatability of the product through proper preparation techniques.

Customers expect a certain standard when they dine out. That standard differs from one type of operation to the next. People dining in a fast food, low-priced steak house do not expect the same palatability levels in the meal as they would expect in a high-priced, full-service restaurant. They do expect suitable levels of palatability in each instance, however.

The quality grading system discussed in the previous chapter gives the food service manager some indication of products' palatability. It is necessary, however, to have a much greater understanding of the components that make up palatability and how those components can be altered. This information is increasingly important today for several reasons.

CUSTOMER PRICE RESISTANCE The price of USDA Prime and Choice strip loins and tenderloins is reaching levels that are forcing some restaurants to look for alternatives. Lower grade strip loins and tenderloins may be substituted if acceptable palatability can be attained. Other cuts from the beef carcass can be merchandised in place of these higher priced cuts, but the food service manager must know how to tenderize the meat to make it an acceptable menu item.

The USDA's grading changes have meant an effect on the palatability of the less expensive cuts of beef. The reduced level of marbling not only reduces flavor and juiciness, but it may also increase cooking losses and have even greater effect on palatability.

Many new meat products have had a number of their palatability characteristics altered by mechanical or chemical means. Food service managers should be able to evaluate the effects of these alterations to decide whether or not the product is suitable for their particular markets.

A thorough understanding of the factors that influence palatability will aid the food service manager in menu planning, purchasing, and in selecting appropriate cooking methods.

4.1 THE FACTORS THAT INFLUENCE PALATABILITY

Palatability factors may be classified as either perceived or experienced. **Appearance** is the most important perceived palatability factor. Before the guest actually samples a steak or a roast, he or she makes a visual judgment as to how good that product is going to taste. The appearance of the food can tremendously affect the guest's sensual experience of eating the product. **Aroma** is also a perceived palatability factor since it can influence the customer's judgment prior to tasting the product. Aroma is also considered a precursor of flavor.

The factors normally considered to affect palatability are the ones experienced upon eating the meat: **tenderness, juiciness,** and **flavor.** Of these, tenderness has been shown to be the single most important factor to influence the customer's opinion of the palatability of the product.[1]

Before judging flavor or juiciness, the consumer evaluates tenderness. A product may well meet the expected criteria of all the other palatability factors, but if it is rated tough by the consumer then it will be rejected. Since this is the number one palatability concern of the consumer, it is important that the food service operator have the technical background required to ensure tenderness in the products he serves.

4.2 METHODS OF MEASURING TENDERNESS

Mechanical Devices

Researchers measuring tenderness of meat objectively most commonly use the **Warner-Bratzler** shearing device. A core sample, extracted from the product being tested, is placed on the device. A blade is then used to shear the core sample, and the amount of pressure required during the shearing process is recorded on a scale that registers pounds per square inch. The force recorded for any given sample can be compared to a standard product such as tenderloin, that is known to be tender. Because this device is mechanical, it removes the human problem of individual subjectivity. It is interesting to note, however, that there is generally a high correlation between results found by this method and results found when a taste panel is used for the same product.

In 1972, Armour and Company developed the **Armour Tenderometer**, another mechanical device to evaluate tenderness. This device inserts probes into the rib eye muscle at the thirteenth rib. (You will recall that the USDA grader uses the same muscle as an indicator of the quality grade.) The amount of force that is required for the probes to separate the muscle tissue in the rib eye muscle is recorded on a gauge. Armour selects carcasses based on this reading and guarantees their tenderness in what they call the Test Tender Program.

Taste Panel Evaluation

The objective methods used for predicting tenderness are fine for research purposes. Certainly the food service operator should be familiar

Figure 4.1 Warner-Bratzler shearing device.

with the mechanical testing devices in order to benefit from the research that has been done in the scientific community. However, the taste panel test performed under controlled conditions with trained individuals is probably the most reliable method of rating tenderness and is certainly the easiest method for the food service operator to understand and use. The panel may consist of a small group of highly specialized individuals who have been trained over a period of years to sample various food products. Or a panel may be made up of a large cross section of consumers with very little special training other than basic instructions on how to use the score card provided. No matter what type of panel is used, it is extremely important that the testing conditions be controlled. The size of the sample of meat, the temperature of cooking, and the degree of doneness should all be kept constant in order to get an accurate rating on tenderness. This type of testing can and should be done by any food facility that is considering placing a new product on the menu. A group of employees and customers may be used to evaluate a new product entering the marketplace. Table 4.1 illustrates a simple test form panel members could be asked to mark.

It is important to limit the number of characteristics judged during a taste test. This is particularly true with an untrained panel. Do not attempt to obtain a rating on texture, color, juiciness, and flavor all at one time. Even a trained panel becomes fatigued and produces less accurate results if too much is asked of it.

TABLE 4.1   A BASIC TASTE TEST FORM

	Tenderness	Flavor
Instructions: Circle a number from 1 (least desirable) to 5 (most desirable).		
Sample A	1 2 3 4 5	1 2 3 4 5
Sample B	1 2 3 4 5	1 2 3 4 5
Sample C	1 2 3 4 5	1 2 3 4 5
Sample D	1 2 3 4 5	1 2 3 4 5

4.3   THE COMPONENTS OF MEAT PRODUCTS THAT AFFECT TENDERNESS

All meat cuts contain a certain percentage of moisture, fat, protein, and ash. Of these four components, the protein is the one that creates the problem of toughness.

The proteins having the greatest effect on tenderness are the myofibrillar proteins of actin and myosin and the connective tissue (stromal) proteins of elastin and collagen (see Chapter 2). Both the coarseness of the striated muscle and the amount of the connective tissue within and surrounding the muscle play an important role in the palatability of the product. Muscles that are fine-grained and have a small percentage of connective tissue tend to be very tender (for example, tenderloin muscle or strip loin muscle). However, muscles from the chuck, which are much coarser in texture and have a higher percentage of connective tissue, tend to be less tender—that is, tough. Besides the part of animal, other factors affecting tenderness are as follows:

Antemortem
- Physiology of the breed
- Age of the animal
- Type of feed and management

Postmortem
- Cold shortening (muscle contraction)
- Time and temperature of storage—aging
- Mechanical tenderization
- Chemical tenderization
- Cutting methods
- Cooking methods

All of the factors listed have an important effect on the tenderness of meat products. Even though the antemortem factors are out of one's direct control as a food service operator, one should be aware of them. To some extent one has control of these factors by selecting USDA graded meat. The federal grader does in fact take many of the antemortem aspects into consideration when examining the carcass for grading purposes.

Breed, Age, and Feed

Different breeds of animals, intended for different purposes, are usually not raised the same way. Therefore, the breed of an animal generally has some effect on the level of tenderness of its meat. For example, meat from dairy cows used for milk production is generally not as tender as meat from steer or heifer raised for beef.

The age of the animal is a very important determinant of tenderness. As a rule, the younger the animal, the more tender the muscles. As the animal increases in age and uses its muscles more often, two things happen: the muscle becomes coarser in texture and the connective tissue undergoes a chemical change (increased **cross-linkage**) that causes it to be less heat-liable or subject to breakdown of collagen. In cuts from older animals less collagen hydrolizes to form gelatin during cooking. As a result of these two changes, meat cut from an older animal is not as tender.

The feed and management practices used in raising the livestock also affect tenderness. **Grass-fed** cattle left to roam on pasture land to obtain feed will use their muscles more and will generally require a longer period of time to reach market weight. Consequently, they will be less tender than cattle put into feed lots where movement is restricted and a grain ratio is supplied to produce market weight at an earlier age. Smith et al.[2] reported significantly lower palatability ratings and higher Warner–Bratzler shear values for grass-fed steers than for **grain-fed steers** but stated that the palatability differences were eliminated when grass-fed steers were fed a high protein diet for 49 days before slaughter.

Cold Shortening

Of the post mortem factors, the food service operator has control over all but the first one. **Cold shortening** is a contraction of the muscle proteins during the initial chilling of the carcass after slaughter. When the animal is slaughtered, the internal temperature rises to about 104°F. In the past, most purveyors concerned with minimizing bacterial growth on surface of the carcass placed the carcass in chilling rooms operating between 28°F and 34°F. Chilling this rapidly within the first 16 hours after slaughter caused a contraction of the myofibrillar proteins and a considerable toughening of the muscle.[3] Locker et al.[4] reported that it was possible that commercial chilling practices caused some meat to become four times as tough because of cold shortening and that the effect was irreversible.

The phenomenon of cold shortening is being dealt with by the purveyors, who are using new systems to prevent or lessen the possibility of cold shortening. Swift and Company has received an approval to use a process called Chlor-chil involving a water spray that contains 100 parts per million of chlorine. The application of this mixture to the carcass during the first four to eight hours after slaughter reduces the bacteria count by 99 percent.[5] Other chemical solutions, such as acetic acid, have produced similar results.

With reduction of the bacteria count, it is possible to chill the car-

Figure 4.2   The effect of aging on the rib muscle.

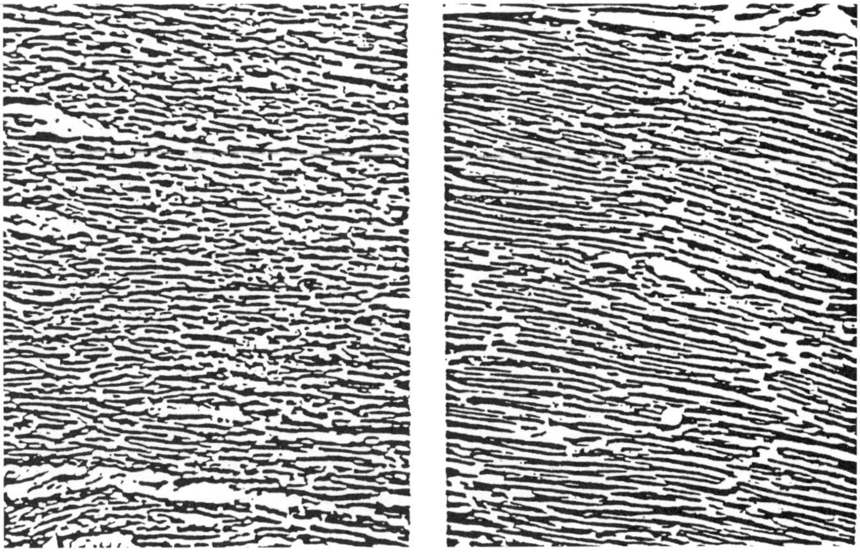

Source: J.F. Price and B.S. Schweigert, *The Science of Meat and Meat Products,* 2nd ed., (San Francisco: W.H. Freeman, 1971), p. 60.

cass at a much slower rate and as a result prevent the onset of cold shortening and the concomitant toughening effect on the muscle. Although food service operators do not have direct control over this aspect of toughening, they should be aware of the problem because it may be one of the factors that could cause even a USDA Prime or Choice strip or rib to be tough.

Time and Temperature of Storage—Aging

Aging is sometimes referred to as natural tenderization, because enzymes already present in meat can, in time, break down the myofibrillar proteins and the connective proteins.

Figure 4.2 shows the effect aging can have on the rib muscle. In the rib eye on the left, which has been aged for 14 days, one can see structural breaks throughout the striation of the muscle. The rib eye on the right has been aged for a period of 28 days and shows almost complete degradation of the striated muscle. If the process were allowed to continue, the muscle would be completely digested by the enzymes and would lose the structure and texture generally associated with a good quality piece of meat.

Time and temperature are the controlling factors in aging. At 0°F, the enzymes are almost at a standstill. With every 8°F rise in temperature, the enzymatic activity doubles, so that at elevated temperatures it is possible to obtain a very rapid rate of tenderization in a relatively short period of time. With temperatures of 70°F, 80°F, and 90°F, very high levels of enzymatic activity are evident and tenderization occurs very rapidly. At these elevated temperatures, bacterial growth also increases very quickly. Except where there are specially designed aging facilities, meat is aged at the temperature normally found in most walk-in coolers, which is 32–38°F.

Some application of the concept of **accelerated aging** due to high temperatures can be seen in **slow roasting techniques**. Roasting a "steamship round," for example, at 200°F for 8, 10, or maybe 12 hours keeps the interior muscle temperature at elevated levels while the exterior surface is maintained above the danger point for bacterial growth. The amount of tenderization that occurs on the interior of the steamship round during this lengthy period of elevated temperature may be equivalent to more than a week of aging at a temperature of 32°F. The longer the elevated temperature is maintained without reaching the point at which the enzymes are denatured, the more tenderization occurs. (See application of this principle in Chapter 17, Cooking Methodology.)

WHAT OCCURS DURING AGING Both the myofibrillar protein and the connective tissue protein are affected by enzymes during the aging period. Within the first few days post mortem, signs of transverse breaks as well as the disappearance of cross striations are visible in the muscle structure. From the first day of aging, this structural breakdown continues to progress. Of the two connective tissues present, only collagen is degraded during aging of meat. Elastin is left virtually unchanged by the enzymatic activity.

Not all muscles benefit equally from aging. It has been observed that the muscles of the rib and loin section show a greater degree of tenderization from aging than muscles from the round or chuck area. (See Section 5.1, Carcass Breakdown.)

CRITERIA FOR AGING MEAT

Quality Generally, the higher quality grades—Prime, Choice, and Good—are the ones that benefit the most from aging. The lower grades are usually so coarse in texture that they are not appreciably improved by the aging process, and the product is such that it will not hold up for the extended period of time required to tenderize it. Lower quality grades can be tenderized more effectively by mechanical or chemical means.

Distance from the Market Distance should be taken into consideration when setting up timetables for aging. In general, meat products are received by the food service operation between seven and ten days after the animal is slaughtered. Consequently, approximately one week of post mortem aging has occurred prior to storage at the food facility.

Use of the Cut The use of the cut will dictate whether or not it should be aged and how much aging is required. For example, a chuck purchased for dry roasting purposes would definitely benefit from the aging process. If the chuck is going to be used to fabricate ground beef, however, then there would be no reason to age it prior to grinding.

Size of the Cut   The larger the cut, the better it will hold up during the aging process, ideally. Carcasses, primal, and major wholesale cuts can be aged effectively. If surface deterioration occurs, a certain amount of trimming becomes necessary but the remainder of the cut is still in very good condition. A portion-cut item, however, if kept for any extended period of time, will generally begin to rot before thorough aging can occur. Therefore, it is recommended that aging be done on the wholesale cut prior to fabricating the individual portions. (Studies are being done on aging portion cuts in a packaging film that has an oxygen and moisture barrier.)

Packaging versus Dry Aging   If a rib is hung in a cooler without any packaging material, a number of things will occur. First, enzymatic activity will tenderize the product over a period of time at given temperatures. Second, the product will suffer a certain amount of shrink loss because of evaporation of moisture into the atmosphere of the cooler (unless humidity control is perfect). Over an extended period of time, a drying out of the surface will generally occur, permitting certain types of molds to grow on the exterior of the muscle. Using this method of dry aging, the meat will become tender and develop a characteristic aged flavor from the mold growth. If the product is packaged in Cryovac, essentially the same tenderization will occur, but the mold growth, the drying out of the surface, and the evaporation loss will be reduced or eliminated because the product will be in an anaerobic state since the packaging material contains an oxygen barrier as well as a moisture barrier.

Losses during dry aging can exceed 10 percent of the product weight. With proper packaging, weight loss can be reduced to less than 1 percent. The major difference between packaged versus dry aging is the flavor change attributed to the mold growth on the surface of the product. The so-called aged flavor is appreciated by a very small percent of the consumer market and many consumers regard this flavor as undesirable.

Whether or not the Product Is Going To Be Frozen   The product that is going to be frozen should be aged prior to the freezing process. There are a number of reasons why this is true. First, aged meat retains moisture better upon freezing and thawing than unaged meat. Second, if the product is frozen prior to aging, upon thawing a certain amount of drip or serum will accumulate in the bag of the packaged product. If the product is aged in this environment, in many cases, the serum will sour during the aging, producing extremely undesirable odors and flavors within the meat product. (Nonpathogenic bacteria present on the surface of the meat cause the serum or drip to produce an acid environment which in turn causes undesirable odors and flavors.) When aging meat that is to be frozen, it is recommended that the amount of time normally used for aging be cut in half to allow for some aging during storage period later, when the frozen product is being thawed.

Costs Involved   Each method of tenderization has a trade-off. The tenderness of the product is increased by aging, but there are some risks involved. Shrink loss and trim loss must be considered in dry aging. Aging after packaging minimizes the shrink loss, but there is still the cost of packaging and storage to consider. Even with these costs in mind, natural tenderization or aging is probably the best method of tenderizing many of the products used in the various food service operations.

LENGTH OF AGING

Beef   Two to three weeks is a substantial amount of time to age wholesale cuts of beef. Most studies show that beyond three weeks the advantages obtained by aging are outweighed by the disadvantages. The following beef cuts from Prime, Choice, and Good carcasses (see Chapter 5, Beef) benefit from a two- to three-week aging period if they are to be used for dry roasts and steaks:

1.   Whole chucks (If the muscles from the chuck are to be used for pot roasts, there is no need to age the product because the cooking method will tenderize the cut.)
2.   Oven-prepared ribs or rib eyes.
3.   Whole strip loins.
4.   Short loins.
5.   Whole sirloins.
6.   Top sirloin butts.
7.   Bottom sirloin butts.
8.   Top rounds.
9.   Sirloin tips from the round.

Lamb   Wholesale cuts from lamb carcasses require only one to two weeks aging time. Domestic lamb is slaughtered at a relatively young age and therefore is quite tender. The cuts that benefit the most from aging are those used for dry roasting and lamb chops. The shoulder, hotel rack, the loin, and the leg section (see Chapter 7, Lamb) will benefit from aging.

Veal and Pork   Veal and pork require little or no aging. The amount of time between slaughter and delivery to the food service facility generally is more than adequate to age the meat. In fact, aging should be avoided for these meats. Because veal has a very high moisture content, it can suffer a tremendous shrink loss during any extended holding period. Pork tends to develop off flavors and odors because of breakdown of its fats as it ages.

Mechanical Tenderization

Tenderization can also be accomplished by mechanically shortening the length of the muscle fiber, as by **scoring** a flank steak, **pounding** a veal cutlet, **grinding** the chuck into ground beef, putting a steak from the bottom round through the **cuber tenderizer**, or using the **blade tenderizing device**.

Mechanical methods, with the exception of one, tend to alter

Figure 4.3 Jaccard blade tenderizer.

dramatically the consistency and appearance of the product. The exception is the blade tenderizing device, which does not appreciably affect the texture and look of the meat. Although when the product is in the raw state, very fine needle holes can be seen on its surface, in the cooked state it is very difficult to notice these fine mechanical breaks. See Figure 4.3 for an illustration of such a device.

The blade tenderizing device, sometimes called a **pinning machine**, makes use of a series of tiny needles that penetrate the muscle fiber, shortening the length of the striation. This device can be used on boneless as well as bone-in wholesale cuts. The blades should enter the cut perpendicular to the muscle grain, thereby shortening the length of the tissue that has to be chewed by the consumer. The advantage to this system is that it is instantaneous and does not produce any flavor changes in the product. It also eliminates the storage or holding time that would be required for natural aging. The trade-off in mechanical tenderization occurs when the drip loss and cooking loss are measured. Structural damage is done to the cell by the pins and as a result studies have shown that there will be approximately 2 percent more drip loss during the storage period and an additional 2 percent shrink loss when the item is cooked. This loss of 4 percent is minimal, however, if the tenderness of the product is upgraded. Nonetheless, in view of this loss, it is not recommended that all cuts be blade tenderized. Prime and Choice strip and rib eye muscles, for example, benefit more from aging and should not be blade tenderized. Lower grade rib eyes and strips can be mechanically tenderized very effectively. Some choice cuts such as the top sirloin butt, the muscles from the bottom sirloin butt and the primal round can benefit from blade tenderization. This is particularly true if these cuts are going to be used for steak purposes. Most purveyors can provide this service for the food operator at a very minimal cost (a few pennies per pound).

Chemical Tenderization

Tenderization by the addition of plant enzymes to meat may accomplish the same task as natural aging but at a much quicker rate because the enzymes are very efficient. The enzymes most commonly used are as follows: **macin** from the Osage orange, **ficin** from the fig, **bromelin** from the pineapple, and—probably the most widely used—**papain** from the papaya. The enzymes are extracted from the raw fruit, dehydrated into a powder form, and usually mixed with a carrier such as salt.

Many food service operators and consumers often voice the concern that if the plant enzyme does such an effective job of tenderizing the meat product it will do damage to the stomach lining when it is ingested. But all of these enzymes are very temperature-specific and pH specific, and hence if the enzyme is not denatured (and so, deactivated) by the heat of the cooking process, it will be denatured by the high acid level (low pH) in the stomach. Papain, for example, is active only between pH 5 and 5.5. Since the pH in a human stomach is between 2 and 3, papain is inactivated there.

THREE METHODS   Swift and Company have a patented process called Swift **Proten Beef**. This process injects an enzyme into the live animal prior to slaughter to allow the circulatory system of the animal to distribute the enzyme to the various muscles. The enzyme is not active while the animal is alive because the pH in the animal's muscles is 7.5, which is too high for it. After the animal is slaughtered, there is a buildup of lactic acid in the muscles due to the breakdown of glycogen. The pH drops to about 5.5, activating the enzyme. At this point tenderization begins.

Post Slaughter Method—Injection   Some processing plants tenderize meat by injecting an enzyme solution under pressure into individual wholesale cuts. The advantage of this method over the Swift Proten method is that it allows the plant the option of regulating the amount of enzyme that goes into any one cut, so that overtenderization of certain sections of the carcass can be avoided. (The tenderloin would not be injected at all, and only small amounts would go into the strip loin and rib section, while the round and chuck would receive a greater amount of the enzyme treatment.)

Dip Method   In another method, the individual portion cut is dipped into an enzyme solution. This can be done either at the plant or at the food service operation. The major disadvantage to this particular method is that only minimal penetration is obtained. Therefore, surface tenderization occurs while the center of the product is relatively tough. Cuts less than ½ inch thick can be effectively tenderized with this method, however.

All of the methods described for chemical treatment are relatively fast and inexpensive ways to tenderize meat products. The enzymes from the various plant sources require a certain amount of time at a given temperature in order to effectively tenderize a cut of meat. Some of the enzymes are more efficient on the muscle protein than on the connective tissue protein, and it is for this reason that many commercial preparations contain a mixture of more than one of these plant enzymes.

A food service operator using a chemically treated product must take the necessary time to instruct the production staff in the proper handling of these items. Manufacturer's recommendations must be followed very closely in order for good results to be obtained. If the enzyme is not given sufficient time to work, the product will be tough. On the other hand, if too much time is allowed to pass, the product may be overtenderized with a mushy, undesirable texture as the result.

The most serious disadvantage to chemical treatment is that the enzymes tend to produce certain off flavors when the product is cooked beyond medium in doneness. This flavor has been described as a liverish taste in well-done products. It is possible to mask this flavor by the use of seasoned marinades. A popular marinade used by many food operators is a teriyaki base made with soy sauce.

WHAT CUTS SHOULD BE CHEMICALLY TENDERIZED
Chemical treatment is used on the lower grades from more mature animals and on less tender cuts from higher grades that one wishes to upgrade in tenderness. Some examples are

1. USDA Standard (mature), Commercial, or Utility strip loins, ribs, and sirloin cuts
2. Less tender cuts from Choice and Good grades—chuck, shoulder clod and bottom round (when using them to produce steaks)

Cutting Methods

The method of fabrication can also be considered as a way of tenderizing a product. Cutting steaks or carving roasts across the grain produces the tenderest product because it gives the customers the shortest length of muscle striation to chew through. Removing connective tissue from the exterior of the muscle also yields a product of increased tenderness.

Cooking Methods

Selecting the proper cooking method for a given cut can alter tenderness greatly. Cuts with high percentages of connective tissue are often cooked with moist heat for a long period of time to allow the collagen present in the muscle to soften and form gelatin. Dry roasted items can have substantial increases in tenderness by lowering the oven temperature to allow for an increase in enzymatic activity over a longer period of time. High-quality cuts from the rib and loin used for steaks are generally cooked with relatively high heat for a short period of time because there is less concern with connective tissue than there is with toughening the actual muscle itself.

Other Methods Used to Increase Tenderness

Recent studies have shown that the normal method of hanging beef carcasses from the **Achilles tendon** (see chart of beef skeleton in Chapter 4, Beef) actually causes a contraction of the **longissimus dorsi** muscle (rib and loin) and, as a result, decreases the tenderness. It has been suggested that a number of methods could be used to prevent this toughening effect. One such method is to hang the carcass from the **aitch bone** to prevent contraction of the back muscle. The Department of Animal Science at Cornell University has developed a procedure to prevent toughening called the **Cornell Stretch Tender Method**, which uses a series of telescoping rods that hold the back muscles in an extended state while the carcass goes through rigor. Both of these methods have proven to be successful in increasing tenderness.

A relatively new method being tested by various slaughter operators is one of **electrical stimulation**. Electrodes are attached to a side of beef and 13 impulses of four to eight seconds duration are used to stimulate the muscles. Current studies show that the greatest benefit of electrical stimulation is produced on lower quality, less tender carcasses and that carcasses that are inherently tender are improved to a much lesser extent.[7]

MARINATION The use of mild acids such as wine, vinegar, or lemon juice has generally been considered to have minimal effect on the tenderness of the product marinated in these solutions. A recent study by Wenham and Locker[8] demonstrates that more than just the acid itself may actually alter tenderness in the product. If the pH of the product is reduced to between 4 and 4.5, the acid environment created seems to stimulate the action of cathepsins, a class of natural occurring proteolytic (protein-attacking) enzymes, which degrade both muscle and connective tissue protein. In order for this to occur, the items being marinated must be relatively thin or must be marinated for an extensive period of time in order to have thorough penetration beyond the surface.

The cuts benefiting most from marination are those with a higher percentage of connective tissue. Cuts with lower amounts of connective tissue are not appreciably affected and the major advantage to marinating these cuts seems to be one of flavor enhancement.

4.4 OTHER PALATABILITY FACTORS

Color

Color has been listed as a perceived palatability factor because the initial response of the customer to the appearance of the product may signal approval or disapproval. The color of cooked meat depends on a number of factors: the amount of myoglobin in the raw product, the type of cooking method used, the temperature used during the cooking process and the internal temperature or degree of doneness reached in the end product.

Generally speaking, muscles with higher amounts of myoglobin in the raw state have a deeper color in the cooked state.[9]

Muscles cooked with moist heat show little surface browning and a dull gray interior, whereas steaks and roasts prepared with dry heat and higher temperatures have a uniformly brown surface due to the denaturation of heme compounds as well as the decomposition and polymerization of carbohydrates, fats, and proteins. The interior appearance of a steak or roast will be affected by the internal temperature reached. A roast brought to an internal temperature of 115°F will show a bright red color, a characteristic of meat prepared rare. From 130°F to 145°F the interior of the meat would be pink and classified medium, and at about 150°–160°F the interior would be gray and classified as well done. (See Figure 3.5, Degrees of doneness, in Chapter 3.)

Juiciness

When one bites into a cut of meat, there are two separate stages of juiciness realized. The first stage is the initial wetness caused primarily by the moisture level within the product. This sensation disappears quite rapidly while the more long-lasting feeling of juiciness is caused by the intermuscular fat present in the meat. A good example of this can be found by comparing veal to mature beef. Following the initial bite into a veal cutlet, there is a large percentage of moisture making the product appear juicy. This dissipates very quickly, however, and the product is found to be actually quite dry in texture. In mature beef with a good amount of marbling, the juiciness level is sustained by the extraction of fat as one chews. Time, temperature, and cooking method have a great effect on the juiciness of a given product.

Selecting the cooking method that will retain the highest amount of moisture and fat will produce the juiciest end product. For example, a roast cooked at a low oven temperature to a rare degree of doneness will be much more juicy than the same roast cooked at a high temperature until well done.

Flavor and Aroma

Lean meat (with fat removed) whether from veal, beef, pork, or lamb exhibits a meaty aroma and flavor when heated. It is believed that aroma and flavor result from a nonenzymatic, browning reaction between free amino acids and reducing sugars during cooking.[10] Since beef, veal, pork, and lamb exhibit similar amino acid compounds upon heating, they all yield a meaty flavor and aroma.

The characteristic flavor and aroma of the species which tells you that you are eating lamb, not beef, for example—can be attributed to the fat. Upon heating, volatile constituents are given off by breakdown of fatty acids and by other compounds stored in fat. Studies performed mixing lean meat from one species with the fat from another species showed that the lean meat took on the characteristic odor and flavor of the species from which the fat came.

A practical application of this information might be the formulation of a chicken-beef hamburger that utilizes chicken meat from spent-layers and beef fat to produce a lower-cost yet palatable burger. More research into the flavor and aroma of meat is definitely needed. The results of such research would be valuable for production of new meat products and meat analogs.

4.5 SUMMARY

The food service manager should be familiar with all factors that influence the palatibility of meat. The control and manipulation of these factors in the storage and preparation of meat is essential if the guest is to be served a product with the best possible level of tenderness, juiciness, aroma, and flavor.

The information provided in this chapter should prove useful to the food service manager when evaluating the suitability of a new product or process. The technical information present will be applied in each of the product chapters and again in the chapters on purchasing, receiving, storage, and preparation.

CASE STUDY

1. A well-established steak house prides itself on serving only USDA Prime strip loin steaks. One evening the restaurant's manager is summoned to the dining room. A customer complains that his steak is as tough as shoe leather. The manager is astonished and states, "We only serve U.S. Prime beef. It must be tender!" The customer replies, "Try it. You won't like it!" The manager takes a taste and, sure enough, the meat is very tough and chewy. He apologizes and orders another steak for the guest. The manager is confused. He cannot understand how this could happen.

   Using the information you have gained from the text thus far, give him several explanations for this occurrence.

2. A family restaurant is faced with a volume/profit squeeze. Their most popular dinner steaks have been USDA Choice New York strip loin steaks and filet mignons. With the increase in beef prices their costs have jumped dramatically. The owner of this food service facility has little technical background. As a result, he sees only two alternatives: increase prices or cut portion sizes to maintain his profit margins. He realizes that either of these alternatives will meet with customer resistance and his volume may drop. If this occurs he will once again be in an unprofitable situation, but he sees no other solutions.

Using the technical information provided in this chapter, pose several solutions to this dilemma. You may wish to refer to the next chapter, on beef, for additional information.

I REFERENCES

1. J.F. Price and B.S. Schweigert, *The Science of Meat and Meat Products,* 2nd ed. (San Francisco: W.H. Freeman, 1971), p. 329.

2. G.M. Smith and Others, "Influence of Feeding Regime and Biological Type on Growth, Composition and Palatability On Steers," *Journal of Animal Science* (August 1977), vol. 45:236–53.

3. B.B. Marsh, P.R. Woodhams, and N.G. Leet, "Studies in Meat Tenderness. The Effects of Tenderness of Carcass Cooling and Freezing Before the Completion of Rigor Mortis," *Journal of Food Science* (1968), 33:12.

4. R.H. Locker, G.J. Daines, "Cooking Loss in Beef. The Effect of Cold Shortening, Searing and Rate of Heating, Time Course and Histology of Changes During Cooking." *Journal of Science and Food Agriculture* (October 1974), 1239–1248.

5. Edward F. Heitter. Paper presented at the Meat Industry Research Conference. March 20, 1975.

6. Price and Schweigert, 1971, p. 186.

7. G.C. Smith, T.R. Dutson, and Z.L. Carpenter, "Tenderize Meat with Electrical Stimulation," *Food Engineering* (August 1977).

8. L.K. Wenhan and Ronald H. Locker, 1976, "The Effect of Marinating on Beef," *Journal of the Science of Food and Agriculture,* 27:1079–1084.

9. Price and Schweigert, 1971, p. 335.

10. Price and Schweigert, 1971.

5 | BEEF

Beef produced primarily for meat consumption is generally raised in three distinct stages: 1) as calf with cow 2) as weaned calf, and 3) in feed lots. In the first stage the beef cow is bred and delivers a calf. The calf remains with the cow for approximately six months and weighs 300 to 350 pounds. At the age of six months the calf is weaned and put out to pasture for an additional six to eight months. At twelve to fourteen months the calf weighs 700 to 750 pounds and is ready for the final stage of fattening. Calves that are to yield top-quality beef are put into feed lots and fed a high-protein mixture of corn, soybeans and other ingredients. It takes between three and six more months to produce a live weight of 1000 to 1100 pounds. Cattle held in feed lots for this long generally are graded Prime or Choice and possess the desirable palatability characteristics associated with those quality levels.

Beef left on pasture (grass fed) takes longer to reach slaughter weight and generally does not possess the high quality attributes of cattle finished in feed lots. Grass-fed beef is leaner and can be less tender than grain-fed beef. It can be very acceptable when used for roasts, pot roasts, stews, and ground beef, however. Some sections of the carcass may also be used for steaks when proper tenderization methods are applied.

5.1 BEEF CARCASS BREAKDOWN

Beef, the most popular meat, appears in many forms on the menu. The beef carcass is the source of highly desirable roasts, steaks, stew meat, and ground beef. Because of the large variation in quality of the different sections of the carcass, the price varies considerably for cuts from different sections.

Purchasing agent, receiving clerk, chef, and menu planner all must know 1) how to identify each wholesale and portion cut, 2) how to rate the quality of each cut, 3) the best way to use each cut, and 4) the market value of each cut. With this knowledge, they can make accurate decisions in developing, pricing, and producing profitable menu items.

This chapter provides numerous illustrations and vital information about how to identify, use, and tenderize primal, wholesale, and portion-controlled cuts of beef. The carcass is the starting point. It is important to know where the wholesale cuts come from on the carcass in order to understand the relative value of those cuts and how they can be used. Cuts from the chuck, which is the shoulder area, are generally the least valuable because of coarseness of the muscle and the high percentage of connective tissue. The back section, which includes the rib, short loin, and sirloin, yields the most desirable and expensive portions. Because these muscles are used for support rather than locomotion, they are the most tender and relatively free of connective tissue. The round differs substantially from one muscle to another; in general, it may be classified as between the chuck and the loin section in palatability. Some very good steaks and excellent roasts can be obtained from the round.

Figures 5.1 and 5.2 show the location of each wholesale cut on the carcass. They also show the bone structure, which is necessary to know since the bone can be used to identify each cut after it has been removed from the carcass.

Figure 5.1   Beef primal cuts.

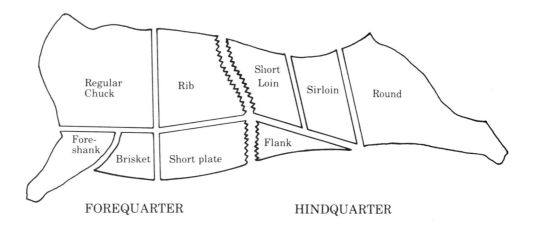

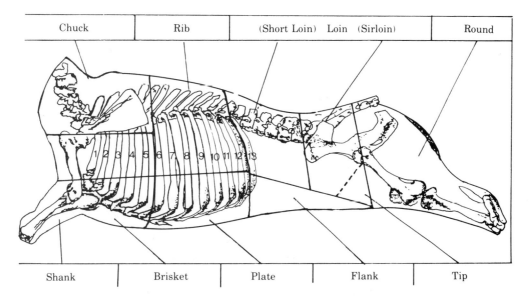

Figure 5.2 Skeleton of beef.

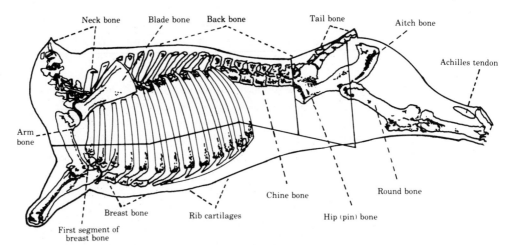

Figure 5.3 shows the breakdown of the live animal, and Table 5.1 gives the percentage yields for each primal and major wholesale cut. Figure 5.4 provides a further breakdown of the primal and wholesale cuts from the beef forequarter, and Figures 5.5 and 5.6 provide the same breakdown for the full loin and primal round, respectively. The remainder of the chapter discusses the major wholesale and portion cuts obtained from the beef carcass, explaining how they are identified, tenderized, and best utilized in a food service operation.

Figure 5.3 Breakdown of beef carcass.

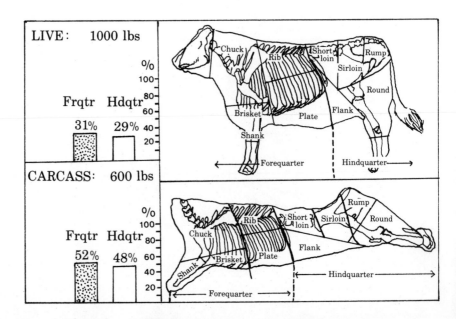

TABLE 5.1 PRIMAL AND WHOLESALE CUTS FROM BEEF CARCASS

Forequarter	Proportion of the Carcass (Approximate Yield, %)
Square-cut chuck	26.8
Primal rib	9.6
Shank	3.1
Brisket	3.8
Plate	8.3
Total	51.6
Hindquarter	
Full loin	17.2
Flank	5.2
Round	22.4
Misc. (kidney, hanging tender, fat, suet, cutting losses)	3.6
Total	48.4

Note: The numbers included in the names of certain cuts are standard identifying numbers assigned by the National Association of Meat Purveyors (NAMP) and listed in their *Meat Buyers Guide.* (See Chapter 14, Purchasing, for more on this numbering system, which identifies primal, wholesale, and portion cuts.)

Figure 5.4   Cuts from the forequarter

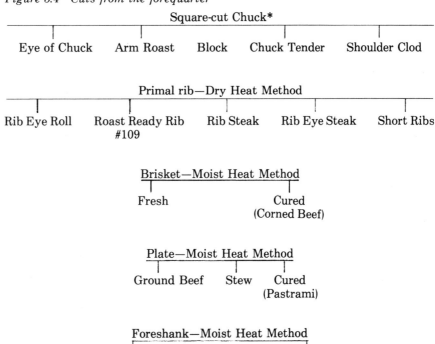

* Dry or moist heat method of preparation depending on quality. Steaks can be fabricated from chuck if mechanically or chemically tenderized.

5.2 SQUARE-CUT CHUCK
—A PRIMAL CUT

Identification
1. Includes 5 rib bones (1st–5th).
2. One side exposes the shoulder blade bone, and the other side exposes the arm bone.
3. Largest cut of the forequarter.
4. Approximate weight (range B) 79–93 lb.

Utilization This primal cut yields roasts, stew meat, and ground beef, as described in the following sections on wholesale cuts and their utilization.

Chuck Roll (or Eye of Chuck)

Identification
1. Large, boneless roast, with no fat cover. One face looks like the rib, the other looks like the neck.
2. Large percentage of connective tissue.
3. Approximate weight (range B) 15–18 lb.

Utilization The best cut within the chuck, the chuck roll can be used for dry roasting if it is of USDA Prime or Choice quality. It also makes an excellent moist heat roast. The first few cuts from the 5th, 4th, and 3rd rib make good broiling steaks if from Prime or Choice carcasses.

Arm and Block

a. Block　　　　　　　　　　　　　　　　*b. Arm*

Identification
1. Arm bone is visible where it attaches to the shoulder blade and where it attaches to the shank.
2. 5 rib bones are attached to the underside of the cut.

Utilization   Two good roasts for dry roasting can be obtained from this section. The better roast would be from the block just behind the arm bone. The other would be the muscle on the front side of the arm bone. The two cuts are not well utilized by food service operators, but when dry roasted slowly, they make excellent roasts for sandwich beef, particularly when sliced thin on an electric slicer. These cuts are generally much less expensive than cuts from the round.

Other useful by-products from this wholesale cut are the short ribs and stew meat obtained from the block section.

Tenderization   Arm and block roasts from Prime and Choice carcasses can be tenderized by aging. They benefit from blade tenderization, having more connective tissue than cuts from the round.

Shoulder Clod

Shoulder clod steaks

Identification
1. Lean, boneless, triangular-shaped cut composed of two intersecting muscles lying above the shoulder blade bone.
2. A thick white collagen connective tissue runs across the surface of the muscle.

Utilization Can be used as a small roast item or cut into lower clod is generally used for moist heat preparation. With the connective tissue cut out, the two individual muscles can be used as minute steaks, cut into cubes for beef kebobs, or used for braised Swiss steaks.

Tenderization Mechanical tenderization by blade tenderizing can be helpful, but because the two muscles run in different directions, only one is effectively tenderized by blade tenderization. Sometimes a combination of chemical treatment and mechanical treatment gives an acceptably tender result.

Chuck Tender

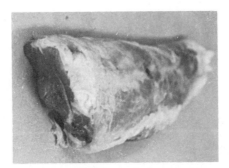

Identification Resembles the tenderloin but has more connective tissue.

Utilization Can be used as a small roast item or cut into lower quality steaks. Although these steaks resemble a filet mignon, they will be substantially less tender.

Tenderization Aging, mechanical tenderization, chemical tenderization, or some combination of treatment may be effective on this cut depending on its quality and intended use.

5.3 PRIMAL RIB, #103

Loin end

Chuck end

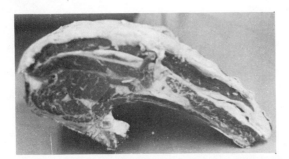

Identification
1. Includes 7 rib bones (6th–12th) and chine bone.
2. Chuck end exposes the shoulder blade cartilage, backstrap, and exterior cap meat over the major rib eye muscle.
3. Approximate weight 28–36 lb.

Utilization   The primal rib yields the most valuable wholesale and portioned cuts from the forequarter area. The breakdown products are described in the following sections.

Note: Many variations of ribs are fabricated from the primal rib, and each one has a different trim.

Roast Ready Rib, #109

Chuck end (6th rib)   *Loin end (12th rib)*

Identification
1. Includes 7 rib bones (6th–12th).
2. Blade bone, cap meat, and chine bone are removed.
3. Rib bone extends 4 in. below the rib eye muscle at the 6th rib and 3 in. below rib eye muscle at the 12th rib.
4. Fat cover is tied back on the roast.
5. Approximate weight (range B) 16–19 lb.

Utilization   The roast ready rib (#109) is one of the most popular versions for restaurant use. It is used as roast prime ribs.

Tenderization   USDA Prime, Choice, and Good ribs should be aged 2 to 3 weeks. Lower grades may require mechanical tenderization using a blade tenderizing device.

Rib Steak, Bone-in

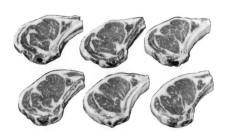

Identification

1. Includes the rib eye muscle and rib bone.
2. Portion size (thickness, weight, trim) varies according to specifications.

Utilization It is the best broiling steak from the forequarter.

Tenderization Age the wholesale cut 2 weeks prior to fabricating steaks.

Rib Eye Roll

Chuck end

Identification

1. Boneless roll with a natural fat cover.
2. Includes the eye muscle of the primal rib.
3. Approximate weight (Range B) 6–8 lb.

Utilization Used as a high-quality roast or to fabricate rib eye steaks.

Tenderization Age USDA Prime and Choice wholesale cuts for 2 weeks. Lower grades may be blade tenderized.

Rib Eye Steak

Identification

1. Rib eye muscle cut from the rib eye roll.
2. Completely boneless.
3. Portion size (thickness, weight, trim) varies according to specifications.

Utilization Considered one of the best broiling steaks from the forequarter.

Tenderization Age the wholesale cut 2 weeks prior to fabricating steaks.

5.4 FORESHANK

Cuts from foreshank   *Foreshank*

Identification   Contains the shank bone and a large percentage of connective tissue.

Utilization   Commonly used for ground beef and stock, the liquid produced by simmering bones in water and used as the basis for soups and sauces.

Caution: Because of extensive amount of connective tissue in the foreshank, it is not recommended that a high percentage of this cut be used in ground beef formulations.

5.5 BRISKET

Identification
1.   Coarse, boneless muscle with a fat cover.
2.   Boneless brisket (with the deckle—or fat layer between brisket and breastbone—removed) weighs approximately (range B) 8–10 lb.

Utilization   Fresh brisket can be boiled or pot roasted. Brisket cured in brine is commonly called corned beef.

Tenderization   Moist heat preparation is necessary to tenderize this cut.

5.6 PLATE

Diaphragm

Identification   The plate is the belly section below the primal rib. It includes 7 lower rib bones, 1 major muscle section, and good fat cover.

Utilization   Commonly used for short ribs, ground beef, and stew meat. The muscle section can also be cured (covered with pepper and spices and smoked after cooking) for pastrami. The diaphragm muscle, which lies on the inside of the plate is cut into steaks called skirt steaks. These are useful as sliced steaks.

Figure 5.5   Breakdown of full loin.

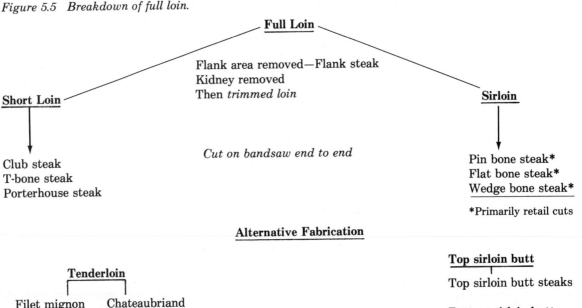

5.7 FULL LOIN – A PRIMAL CUT

Rib end

Sirloin end

Identification
1. Includes 13th rib bone and 6 lumbar and 5 sacral vertebrae.
2. Can be further divided into the short loin and sirloin or into strip loin, tenderloin, top sirloin butt, and bottom sirloin butt.
3. Approximate weight, trimmed (range B) 42–50 lb.

Utilization This primal cut is fabricated into some of the most desirable wholesale cuts for steaks and roasts.

Short Loin

Identification
1. Includes 1 rib bone (the 13th), 6 vertebrae, the strip loin muscle, and the tenderloin muscle.
2. Approximate weight (range B) 21–25 lb.

Utilization This wholesale cut can be used to fabricate the steaks classified as Club, T-bone, and Porterhouse steaks, or it may be divided into its individual muscles: the tenderloin and the strip loin muscle.

Tenderization Aging is the best method of tenderization for the higher grades, that is, Prime or Choice. Lower grades may benefit by blade tenderization.

CLUB STEAK

Identification
1. Located at the forward end of the short loin.
2. Contains loin eye muscle but little or no tenderloin.
3. Smallest steak in the short loin.
4. Portion size (thickness of cut, weight, and trim) varies according to specifications.

Utilization   Steaks of high quality for broiling.

Tenderization   Aging should be done on the wholesale cut prior to fabrication. In the case of lower grades, some may benefit by mechanical blade tenderization.

T-BONE STEAK

Identification
1. Includes the loin eye muscle and a section of the tenderloin muscle.
2. Diameter of the tenderloin must be no less than ½ in.
3. Portion size (thickness, weight, and trim) varies according to specifications.

Utilization   Steaks of high quality are broiled.

Tenderization   Aging should be done on the wholesale cut prior to fabrication. In the case of lower grades, some may benefit by mechanical blade tenderization.

PORTERHOUSE STEAK

Identification
1. Includes the loin eye muscle and a large section of the tenderloin muscle.
2. Diameter of tenderloin must be no less than 1¼ in.
3. Portion size (thickness, weight, and trim) varies according to specifications.

Utilization: Steaks of high quality for broiling.

Tenderization Aging should be done on the wholesale cut prior to fabrication. In the case of lower grades, some may benefit by mechanical blade tenderization.

Full Tenderloin

Regular   *Special*

Identification
1. Regular full tenderloin has fat trimmed to the blue tissue at a point not beyond ¾ in. of the length of the entire tenderloin measured from the butt end and weighs approximately (range B) 5–6 lb.
2. Special full tenderloin has surface fat, side strip muscle, and the fat lying between the side completely removed. Membranous tissue over the tenderloin muscle remains intact. Approximate weight (range B) 3–4 lb.

Utilization The tenderloin yields the tenderest steak: the filet mignon. It may also be cut into double-thick steaks classified as Chateaubriand. It is also cut into sections called medallions or tournedos of beef. The trim meat is used for beef tips, *boeuf en brochette* (skewered beef), and items such as beef stroganoff.

Tenderization This wholesale cut is tender by nature because of its location within the carcass. Aging should be the only method of tenderization used.

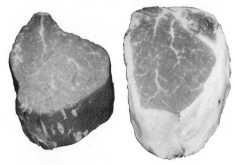

Side muscle off   *Side muscle on*

FILET MIGNON

Identification

1. Boneless cut of beef cut from the tenderloin.
2. Portion size varies according to specifications.
3. Thickness generally 1½ in. to 3 in.

Utilization Although this is the most tender steak from the beef carcass, it tends to lack flavor and is usually broiled and served with a sauce.

Tenderization: Full tenderloin is aged prior to cutting steaks.

Strip Loin

Bone-in   *Boneless*

Identification

1. Bone-in strip loin includes 1 rib bone, featherbones, and loin eye muscle and weighs approximately (range B) 13–16 lb.
2. Boneless strip loin has all bones removed, only loin eye muscle remaining, and weighs approximately (range B) 10–12 lb.

Utilization Primarily used to cut New York strip steaks, either boneless or bone-in. The boneless strip loin may also be used as a very high quality roast item.

Tenderization The top grades, Prime and Choice, should be tenderized by aging the wholesale cut for approximately 2 to 3 weeks. Lower grades may need a blade tenderizer to ensure tenderness.

Strip Steak

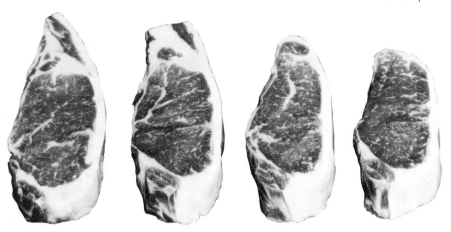

Identification
1. Cut from the boneless or bone-in strip loin.
2. Portion size (thickness, weight, and trim) varies according to specifications.

Utilization   Steaks of high quality are for broiling.

Tenderization   Aging should be done on the wholesale cut prior to fabrication. In the case of lower grades, some may benefit by mechanical blade tenderization.

Sirloin

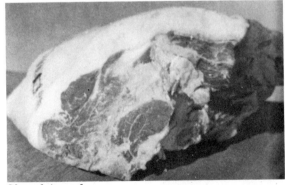
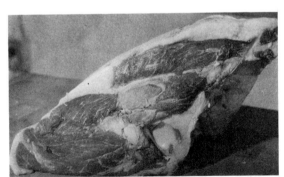

Short loin end

Sirloin end

Identification
1. Includes hip bone, butt tenderloin, and muscles from the top and bottom sirloin butt.
2. Can be further divided into the top sirloin butt and bottom sirloin butt.
3. Approximate weight (range B) 19–24 lb.

Utilization   In food service the sirloin is generally broken down into its components and used as roasts and steaks.

Tenderization   The wholesale cut of the sirloin should be aged approximately 2 weeks prior to fabricating it into its components.

Top Sirloin Butt

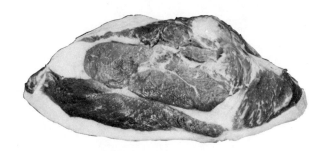

Identification

1. Two major muscles running in opposite directions, separated by a layer of fat and connective tissue.
2. Interior fat globule.
3. Muscle is large at one end and tapers where it joins the strip loin.
4. Approximate weight (range B) 7–9 lb.

Utilization   Can be used as a roast or cut into individual steaks. In many cases the top sirloin butt is divided into its major muscles and then fabricated into top sirloin butt steaks. These steaks have excellent flavor but are somewhat less tender than steaks cut from the strip loin. Separating the muscles improves tenderness because the striation (grain) of each muscle runs in opposite direction.

Tenderization   Top quality (Prime and Choice) top sirloin butts may be effectively tenderized by aging. To ensure tenderness, many operators have even Choice top sirloin butts run through the blade tenderizing device prior to cutting individual portion-cut steaks.

TOP SIRLOIN BUTT STEAKS

Identification
1. Cut from the top sirloin butt.
2. Portion size (weight, thickness, and trim) varies according to specifications.

Utilization   As medium-quality broiling steaks

Tenderization   Wholesale cut should be aged and/or mechanically tenderized before steaking.

Bottom Sirloin Butt

Identification
1. Includes the flap, ball tip, and triangle tip.
2. Two lines of connective tissue run parallel to the top of the muscle.
3. Cap muscle lies above the muscle as viewed from the round end.
4. Approximate weight (range B) 5–6 lb.

Utilization   Generally broken down into its components—the ball tip, triangle tip, and flap—to be used for steaks and beef tips. Many steaks are currently fabricated from this wholesale cut, which resembles some of the higher priced steaks such as filet mignon and strip steak. Because of the small size of these muscles, fabrication is generally carried out more efficiently by the portion-control companies.

Note: Only two to four steaks are obtained from each of the individual muscles within the bottom sirloin butt.

Tenderization   The individual cuts within the bottom sirloin butt are generally blade tenderized prior to fabricating the portion-cut steaks.

Flank Steak

Identification
1. Flat, boneless, oval-shaped muscle with elongated muscle fibers.
2. Practically free of fat.
3. Approximate weight (range B) 1–2 lb.

Utilization Commonly used as London broil or as a sliced steak item.

Tenderization Although flavorful, it lacks tenderness and therefore should be scored to shorten muscle fibers and sliced thinly on the bias to ensure acceptable tenderness.

Figure 5.6 Breakdown of primal round.

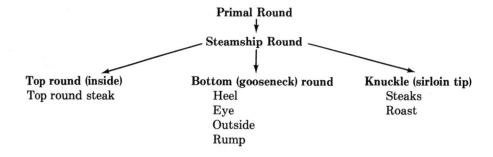

5.8 PRIMAL ROUND

Identification
1. Includes achilles tendon on hindshank, aitch bone, and round bone.
2. Can be further divided into three major cuts: top round (inside), sirloin tip (knuckle), and bottom (gooseneck) round.
3. Approximate weight (range B) 71–83 lb.

Primal Round

Steamship Round

Utilization It is sometimes used as a large steamship round with the shank and aitch bone removed. Normally it is broken down into the individual wholesale cuts and utilized for roasting purposes and to some extent for less expensive steaks.

Tenderization The entire round may be aged for a period of approximately 2 weeks to ensure tenderness of the wholesale cuts.

Knuckle or Sirloin Tip

Identification
1. Triangular cut made in front of the round (femur) bone.
2. Hole located beneath the knuckle where the kneecap was removed.
3. Approximate weight (range B) 9–11 lb.

Utilization This cut is underutilized by most food operations. It is a continuation of the bottom sirloin butt into the round and, if prepared properly, it makes good quality broiling steaks. It also makes an excellent dry roast when roasted slowly and is generally less expensive than the top or inside round.

Tenderization Aging provides an effective way to tenderize this wholesale cut. When using it for steaks, separate the muscles along the connective tissue seams and blade tenderize each individual muscle.

Top Round (inside)

Identification
1. Heart-shaped cut of beef.
2. Good fat cover except for the bald patch that exposes a section of meat.
3. Approximate weight (range B) 17–20 lb.

Utilization The top round is primarily used for roast beef. It can also be cut into fairly thick sections, broiled, and sliced for a sliced steak item similar to London broil. It is also used for cutting beef tips for items such as beef burgundy and lower cost stroganoff.

Tenderization The top round can be aged for a period of 2 weeks prior to preparation. When using it as a roast, use a slow roasting, low-temperature method. The top round may also be blade tenderized when it is being used for individual steaks.

Bottom (Gooseneck) Round

Identification
1. Large, boneless piece of beef with several different muscles; triangular shaped at one end.
2. Can be further broken down into the heel, eye, rump, and outside round.
3. Approximate weight (range B) 21–25 lb.

Utilization It can be utilized as one whole cut but because of differences of muscle structure, is generally broken down into three or four separate muscles. If used as one whole cut, either slow roasting or moist heat preparation is necessary for good results.

Tenderization   Aging can tenderize some of the muscles sufficiently in the gooseneck. The cooking process is also a method of tenderizing this particular cut. The components may be tenderized in a number of ways.

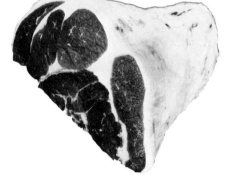

HEEL

Identification
1.   Top, bottom, and eye of round muscles terminate in this cut.
2.   High percentage of connective tissue and very little fat.
3.   Boneless, wedge-shaped cut found in the lower part of the round.

Utilization   Because of the amount of connective tissue and coarseness of the muscle, the heel is generally used for moist heat preparation or ground beef products.

Tenderization   There is no effective way other than grinding to tenderize this product. Moist heat preparation for a long period of time will soften much of the connective tissue but it still will not make a very acceptable roast.

EYE OF THE ROUND

Identification
1.   Tube-shaped muscle with sides tapering to the top.
2.   Approximate weight (range B) 3–5 lb.

Utilization   It can be used for a small oven roast. It may also be cut into sandwich steaks, but these must be tenderized or cooked with moist heat. This cut lacks marbling and is slightly drier than other round roasts.

Tenderization   Aging and/or mechanical tenderization.

OUTSIDE ROUND

Identification
1. Boneless cut of meat remaining after the removal of the heel and eye from the Gooseneck.
2. Muscle is large at one end and tapers down gradually on other side.
3. Approximate weight (range B) 10–13 lb.

Utilization   If dry roasted slowly, it will make an acceptable roast for slicing thin pieces of meat for roast beef sandwiches. It is very often used for moist heat preparation for items such as sauerbraten or pot roasts. The bottom outside round may also be cut into fairly thin steaks and run through a cuber to produce Swiss steak.

Tenderization   Aging, slow roasting, and/or mechanical tenderizing.

RUMP

Identification
1. Triangular-shaped cut found in the round.
2. Front face looks like top sirloin butt and back face looks like outside round.

Utilization   Generally used for small roasts or cut into lower quality steaks.

Tenderization   Aging and/or mechanical tenderization with blade tenderizing device.

5.9 GROUND BEEF, HAMBURGER, CHOPPED CHUCK

Billions of pounds of ground beef are purchased annually by all types of food facilities, including fast food chains, hospitals, schools, and restaurants. As with all other meat products, the standards for ground beef quality vary, and it is up to the food service operator to select the quality and formulation most suitable to his or her own facility's needs. The following information may be useful.

Fat Content

The U.S. Department of Agriculture specifies that anything labeled ground beef, chopped beef, or hamburger should not have more than 30% fat. The range of acceptable fat content is generally 18 to 22%. This amount of fat provides good flavor and juiciness without excess cooking losses. A hospital requiring low-fat hamburger may specify 12 to 15% fat, but the product would be somewhat dry when prepared. In the purchasing specification for ground beef, a lean-to-fat ratio is usually listed. For example, an 80/20 formula has 80% lean meat and 20% fat. Fat analyzers are available that measure fat content accurately within 1 or 2%. These devices are particularly valuable to operators using large quantities of ground beef.

Muscles Used in Formulation

Ground beef can be made from all parts of the carcass, but certain cuts, like the shank and heel, have a high percentage of connective tissue, making them less desirable. If these cuts make up the bulk of the ground beef formulation, the end product will have fragments of connective tissue, which may not soften upon grilling or broiling, leaving the customer displeased as he or she bites into a piece of gristle.

Terms like ground chuck, ground sirloin, or ground round refer to beef ground from those sections of the carcass with no other cuts added. In practice, however, an order for one of these three sometimes produces just a leaner formulation of ground beef.

Number of Grinds

Ground beef for hamburger is generally ground two times to provide uniform particle size and even distribution of fat. If the product is ground too much, heat will develop and the protein in the meat will toughen. (Hamburgers that curl up or form a cup when they are placed on a grill have generally been overheated in grinding.)

Quality of the Meat

It is not necessary to use USDA Choice or Prime beef for hamburger. Lean meat from lower grade beef ground with the trim fat from Choice beef will yield excellent beef for hamburgers, chili, and meat loaf.

5.10   SUMMARY

Approximately fifty different beef cuts have been illustrated in the charts and photographs of this chapter. Many food service operations use only a small number of these cuts. The reasons for using only certain cuts are many, but often the menu planner is simply unfamiliar with the wide variety of carcass cuts and their proper use. With higher costs in general and high beef prices forecast for the future, more operators will seek economic relief by using less expensive beef cuts that can be profitably added to the menu if properly prepared.

This chapter should not only be helpful in selecting the lower cost cuts, but it should also be a valuable tool for identifying products at the receiving dock.

CASE STUDY   The Blank Hotel is a lower-middle-income resort property that is facing a quality/cost/profit squeeze. The typical menu entrees (modified American plan) with the cuts of beef used for each are as follows:

Entrees

Dinner

Roast Beef   prime ribs
London Broil   flank steak
Beef Brochette   tenderloin cubes
Dinner Steak   Porterhouse
Steak Sandwich   New York strip

Luncheon

Pot Roast   bottom round
Hot Roast Beef Sandwich   leftover prime rib
Beef Burgundy   tenderloin tips
Minute Steak   top sirloin butt

Using the information provided by this book thus far, recommend alternative cuts that might be suitable for the menu items listed. To become familiar with the current prices for the existing cuts used and the ones proposed, comparison shop by contacting a local supplier, using the "yellow sheet," or obtaining the *Market Livestock News*, which is published weekly by the U.S. Department of Agriculture. Once the price data are gathered, determine how much could be saved by switching to the cheaper cuts you recommend. (Assume that 100 lb of each product is used per week.)

6 | VEAL

In the past twenty years, dairy herds have been reduced by some 50 percent nationwide. As a consequence, the supply of veal, the flesh of the offspring of dairy cows, has been dwindling. Top quality veal is almost impossible to obtain in retail stores, and much of what is sold as veal is in fact calf. **Veal**, by definition, is a bovine under 3 months of age. It is prized since, because of its youth, the lean has a very fine grained texture, is very tender, and has a delicate, mild flavor. **Calf**, the stage in maturing between veal and beef (3 months to 9 months), lacks the tenderness of veal and lacks the flavor of beef.

When purchasing veal wholesale, one must specify what USDA quality grade is desired—Prime, Choice, Good, Standard, Utility, or Cull. One must also specify what type of veal is desired. Two types are generally available. The first type is **dairy veal**: offspring from dairy herds raised primarily on pasture. This veal has a brownish lean and a smooth textured muscle. A special classification is bob calf, dairy calf only a few days old. **Bob calf** is even more tender than mature veal, but the yield of the carcass is very poor because the muscles are small, not having had an opportunity to develop in size.

The second major type of veal is so-called **nature-fed**. This term is misleading because there are no calves raised commercially for veal that are actually nursed on cow's milk. A more appropriate term would be **formula fed**. These calves, also the offspring of dairy cattle, are confined to stalls to restrict their muscle use. They are fed a milk replacer made from various ingredients, including dry milk solids, soy oil, and whey. Their diet is kept practically free of iron to restrict the amount of myoglobin formation in the muscle and hence produce the desirable milkwhite color of the flesh. Restricting the animals' activity while feeding them constantly produces heavy carcasses with a thin milk-white fat cover and extremely tender muscles. Two well-known brands of this type of veal are **Plume de Veau** from New York and **Provimi** from Wisconsin. First class restaurant operations use formula-fed veal. It costs more per pound than dairy veal, but the yield from each wholesale cut is higher and the overall palatability is far superior. Because this quality is not available to the consumer for retail purchase, it is a good attraction on the restaurant menu.

Figure 6.1 shows the location of each primal and wholesale cut on the veal carcass, and Table 6.1 shows what proportion of the carcass each cut represents. Figure 6.2 shows how the wholesale cuts are further

89

Figure 6.1 Veal primal and wholesale cuts.

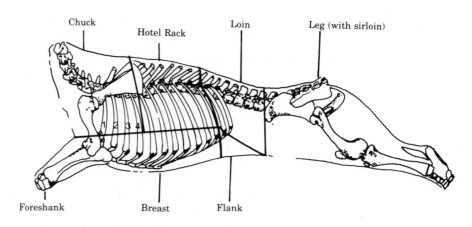

TABLE 6.1   VEAL PRIMAL AND WHOLESALE CUTS

Approximate Yield	%
Chuck	28
Rack	7.3
Loin	7.7
Leg	34.0
Shank	3.8
Breast and flank	13.4
Kidneys and suet	5.8
	100

broken down and utilized in a food service operation. The remainder of this chapter discusses the major cuts obtained from the veal carcass, how they can be identified, and how they can be utilized.

Figure 6.2   Utilization of primal and wholesale cuts of veal.

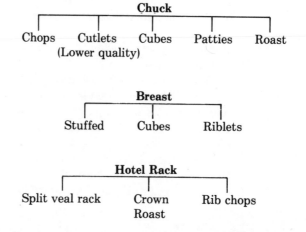

Figure 6.2   Utilization of primal and wholesale cuts of veal. (continued)

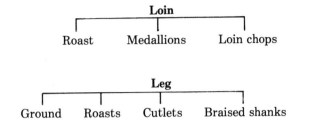

Loin

Roast   Medallions   Loin chops

Leg

Ground   Roasts   Cutlets   Braised shanks

6.1   CHUCK

Single—arm end

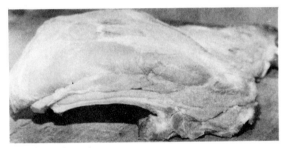

Single—rib end

Identification
1. Each side includes 4 rib bones, a shoulder blade bone, and an arm bone.
2. Corresponds to the square-cut chuck in beef.
3. Approximate weight (range C) 28–36 lb.

Utilization   The chuck can be used in a number of ways: bone-in chops (arm and blade), boneless roasts, veal cubes, and ground trim for veal patties. Lower quality cutlets, which usually require cubing, can also be prepared from the chuck.

6.2   BREAST

Identification
1. Corresponds to the short plate and brisket of beef.

2. Includes breast bone and end of the 11 rib bones.
3. Approximate weight (range C) 8–10 lb.

Utilization Stuffed breast of veal, riblets, or cubes for stew are the major uses of the breast. Moist heat cooking would be an appropriate method of preparation.

6.3 HOTEL RACK, BOTH SIDES

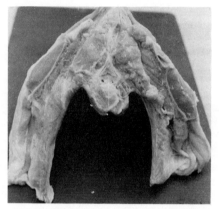
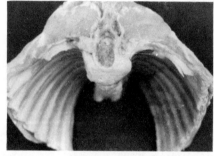

Chuck end   *Loin end*

Identification
1. Corresponds to the primal rib in beef.
2. Each side (veal rack) contains 7 rib bones.
3. Rib bones measure not more than 4 in. from the outer tip of the rib eye muscle.

Utilization The rack can be roasted or broken down into the following cuts: rib chop, split veal rack, or crown roast.

Rib Chop

Identification
1. Includes the rib bone and rib eye muscle and little or no fat covering.
2. Portion size (thickness, weight, and trim) will vary according to specifications.

Utilization Naturally tender, this chop should be pan-fried, braised, or broiled.

Veal Rack

Identification
1. Includes 7 rib bones and shoulder blade bone with little or no fat covering.
2. Approximate weight (range C) 6–7 lb.

Utilization This high-quality roast can be dry roasted slowly to prevent dryness and give adequate tenderness levels.

Crown Roast

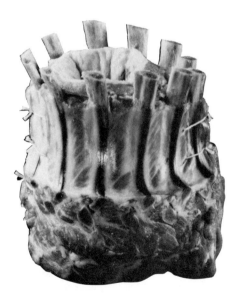

Identification Usually two veal racks tied together, with frenched rib ends (in frenching, 1–2 in. of the rib bones are cleaned of meat).

Utilization. This cut is generally roasted and prepared for service in the dining room.

6.4 LOIN, DOUBLE

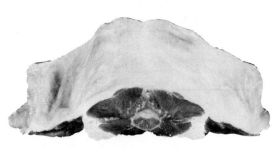

Identification
1. Includes loin eye muscle, tenderloin and two rib bones for each side.
2. Approximate weight (range C), trimmed, 11–14 lb.

Utilization This high-quality cut can be boned, rolled, and tied, then dry roasted or cut into loin chops on the bandsaw. Medallions of veal are sometimes cut from the loin muscle.

Loin Chop

Identification
1. Includes loin eye muscle and tenderloin muscle.
2. Resembles a mini-Porterhouse steak.
3. Portion size (thickness, weight, and trim) varies according to specifications.

Utilization This high-quality chop can be braised, pan-fried, or broiled.

6.5 LEG

Identification
1. Includes the hip bone, aitch bone, femur, and hindshank.
2. Can be further broken down into sirloin area and round (the knuckle, top round, eye and bottom round).
3. Approximate weight (range C) 56–70 lb.

Utilization   The legs produce the highest quality veal cutlets, cut from the individual muscles found in the leg. The cutlets are generally pounded and then prepared for items such as scallopine and veal parmigiana. The legs can also be boned, rolled, and tied for roasting. The individual muscles of the top round, eye, and bottom round are easiest to portion cut into cutlets, whereas the top and bottom sirloin as well as the knuckle make excellent roasts. The heel and hindshank should be used for moist heat preparation or ground. The veal shanks are often cut bone-in and braised for osso bucco, a popular menu item.

Note: When preparing a leg of veal for cutlets, it is important to remove the membrane (connective tissue sheaths) that cover the individual muscles. If this is not done, it will cause the cutlet to curl up, as the connective tissue shortens during sauteing.

Osso Bucco
Slice of Veal Shank

Veal Cutlet
Cubed and Solid

Boneless Veal Leg Roast

6.6 SUMMARY

Veal has often been used as a criterion by which to measure the quality and capabilities of the staff of a restaurant kitchen. It is an item that sometimes distinguishes a truly fine restaurant from its would be competitors.

The fact that veal supplies are declining and prices are climbing can be looked at as both a problem and a blessing. Its price and scarcity have made veal a luxury item that is most often reserved for that special occasion when one dines out. Veal can be used as a drawing card by a knowledgeable restaurateur. The proper selection of a few veal dishes on the menu can add variety, notoriety, and profit.

CASE STUDY Survey ten different types of food service facilities in your area. Study their menu offerings and make a list of the veal dishes that appear on each menu. Try to determine from the menu description what cut of veal is used to prepare the item. After studying these ten menus and considering the different types of food operations you surveyed, propose two menu items (using veal) for each of these operations.

---

7 | LAMB

Less lamb is consumed in the United States than any other meat. Approximately two pounds per person per year on a carcass weight basis were consumed in 1978. Even this small amount of lamb can be a very important menu item to many food facilities, however. Certain lamb cuts—rack of lamb for two and double-cut frenched rib chops—are gourmet items that are usually served only in restaurants. These items command a premium menu price and attract customers because lamb is so seldom served at home.

In food service, lamb competes favorably with some of the higher cost beef items; leg of lamb can compete with ribs of beef, rack of lamb with chateaubriand. Merchandising lamb properly not only attracts new customers, but it also allows some flexibility when other meat items become overpriced during heavy demand periods.

7.1 CLASSIFICATION IS ACCORDING TO AGE

Just as there are different classifications for bovine at various stages of development, the same distinction is made for ovine species. Lamb are young ovine, under one year of age. In the United States most lamb is slaughtered under six months of age. The young animals yield the kind of meat that the average U.S. consumer expects when purchasing lamb: meat that is very tender and mild in flavor. The age of the ovine animal can be verified by checking for the presence of a break joint on the lower front shanks (see Figure 3.9).

At approximately 12 to 14 months of age the break joint does not break because it forms mature bone. As a result the upper spool joint is present. This change is basis for identifying yearling mutton. (Figure 3.15 compares the break joint and spool joint.) With another year increase in age, the classification changes from yearling to mature mutton. Mutton is much less tender than lamb because naturally, the older the animal gets, the less tender the meat becomes. Mutton, moreover, has a strong taste that many American consumers find objectionable.

The most accurate way to determine the age of an ovine is to examine the teeth. Figure 7.1 shows the changes that occur with age.

Lamb mouth with 8 incisors. These temporary teeth are called milk teeth.

Yearling mouth with 1 pair of permanent incisors

Figure 7.1 Changes in teeth of ovine with age.

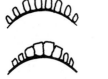

2-year-old mouth with 2 pairs of permanent incisors

3-year-old mouth with 3 pairs of permanent incisors

4-year-old mouth with 4 pairs of permanent incisors

7.2 QUALITY GRADES AND YIELD GRADES

After inspection, the lamb carcass may be graded for quality or yield or both. The characteristics used to assign a quality grade are similar to those used for beef: texture of the lean, color, firmness, and marbling. The quality grades for lamb are Prime, Choice, Good, Utility, and Cull. When the ribbon role stamp is applied to the carcass the term LAMB appears within the grade stamp.

Yield grades are from 1 to 5, where 1 is the ideal and 5 is very wasteful. The factors used to judge the yield grade are carcass weight, amount of fat over the rib, and the conformation score of the legs, that is, amount of muscle development in the leg.

Genuine Spring Lamb

The term **Genuine Spring Lamb** designates lamb marketed between the first Monday in March and the second Monday in October. Some purveyors stamp the term on the carcass. It does not represent a certain quality of lamb. A quality rating is given only by the USDA quality grade stamp.

Low Lamb Supplies

Over the last 30 years the number of lamb and sheep in the United States have been reduced from 52.5 million to 16.5 million in 1974. According to Biglin, there are several reasons for this decline: "(1) less demand for wool, (2) management and herder problems, (3) increased predator problems, (4) competing demands for government owned range land.''[1] Because of increasing demand and dwindling supplies, much lamb is now being imported from New Zealand. The young lamb produced there is very similar in tenderness and flavor to domestic lamb. The major difference is that the New Zealand lamb tends to be a little smaller in size (lighter weight range) than the domestic product.

7.3 LAMB CARCASS BREAKDOWN

Figures 7.2 and 7.3 show the breakdown of the lamb carcass into wholesale cuts, and Table 7.1 shows the proportion of the carcass each primal cut and wholesale cut constitutes. The wholesale cuts are then discussed in detail, with information on identification, tenderization, and best commercial use.

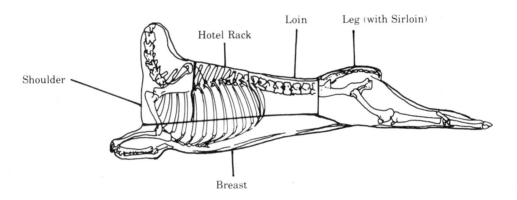

Figure 7.2 Lamb primal and wholesale cuts.

TABLE 7.1   LAMB WHOLESALE AND PRIMAL CUTS

Approximate Yield, %

Foresaddle	
Shoulder	26
Hotel rack	9
Shanks	5
Breast	10
	50
Hindsaddle	
Legs	39
Loin	7
Flanks	2
Kidney and suet	2
	50

Figure 7.3 Utilization of primal and wholesale cuts of lamb.

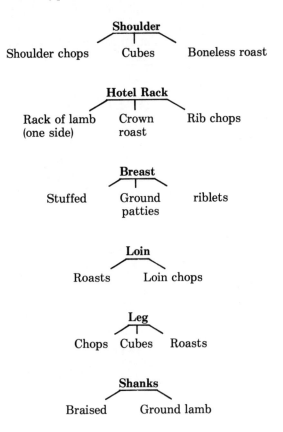

7.4 SHOULDER

Identification
1. Each side includes 4 rib bones, the shoulder blade bone, and the arm bone.
2. Not more than 1 in. of the neck is left on the shoulders.
3. Approximate weight (range C) 13–16 lb.

Utilization When cut on the bandsaw, the shoulder yields shoulder chops (blade and arm), which are generally grilled, braised, or pan-fried. The shoulder can also be boned, rolled, and tied as a roast. Either dry or moist heat are acceptable methods of preparation. The shoulder can be cubed for stew, kebobs, or curries.

Tenderization The shoulder will benefit from 1 week of aging.

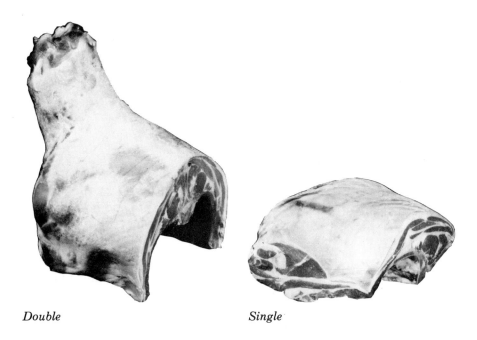

Double   *Single*

Shoulder Chops

7.5 SHANK

Identification
1. Shank bone visible on both sides.
2. If lower shank bone is left on, break joint or spool joint is evident.

Utilization Commonly used for braised lamb shank or ground lamb.

Tenderization Grind or apply moist heat to this cut.

7.6 BREAST (flank on)

Identification
1. Slightly coarse, thin piece of meat.
2. Contains the rib ends of the 12 rib bones and the breast bone.
3. Similar to the short plate and brisket of beef.
4. Approximate weight (range C) 7–9 lb.

Utilization   The breast can be used in a number of ways. One method requires cutting the breast between the lean and ribs (thus forming a pocket), stuffing the pocket, and finally dry roasting the stuffed breast at a low oven temperature. Or the breast can be boned, rolled, and tied, producing a roast for moist or dry heat preparation. If only the breast bone is removed, the breast can be cut between the ribs, yielding an item called riblets to be grilled or braised. The shank, breast, and neck can also be ground and formed into patties.

Tenderization   Moist heat preparation adequately tenderizes the breast. If used for a stuffed breast, dry roast slowly.

7.7 HOTEL RACK

Chuck view

Rib view

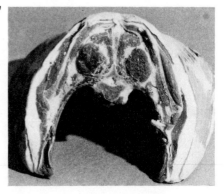

Single rack

Identification
1. One side is called a lamb rack.
2. Includes 8 rib bones and the shoulder blade bone.
3. Ribs measure 4″ from the outer tip of the rib eye muscle.
4. Rib eye muscle is small.
5. Approximate weight (range C) 6–7 lb.

Utilization   A rack can be dry roasted, prepared as a crown roast, or fabricated into rib chops.

Tenderization   Rack should be aged 1 week.

Crown Roast

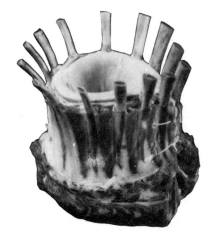

Utilization   Generally, two to three racks are tied together to form a crown roast of lamb. The ribs are frenched and the cut is dry roasted. A crown roast is usually prepared for service in the dining room, where it is carved at tableside or on a buffet line.

Rack Of Lamb For Two

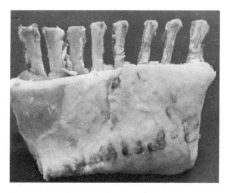

Utilization   Usually roasted and carved for two at tableside.

Rib Chop

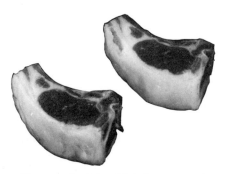

Utilization   Rib chops are generally cut double thick because the area of the rib eye muscle is very small. Dry heat is the best method of preparation—that is, broiling or grilling.

7.8 LOIN, (TRIMMED)

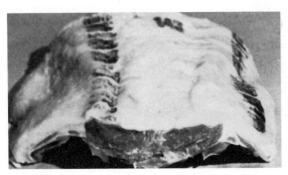

Rib view

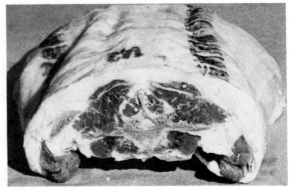

Loin view

Identification

1. Includes the 13th rib bone, the loin muscle, and the tenderloin muscle. (It is equivalent to short loin in beef.)
2. Approximate weight (range C) 5–7 lb.

Utilization This high-quality cut can be boned, rolled, and tied to produce an extremely desirable roast. It is generally cut into loin chops for broiling.

Loin Chops

Identification The first chops from the loin will contain the 13th rib; the remaining chops will contain the loin muscle and varying amounts of the tenderloin muscle. (Similar to club, T-bone, and Porterhouse in beef.)

Utilization This high-quality chop is prepared best by broiling or grilling.

Tenderization Age whole loin 1 week prior to cutting chops.

7.9 LEG

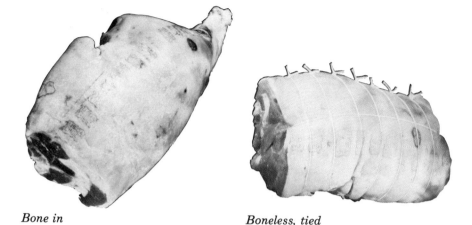

Bone in   *Boneless, tied*

Utilization   The leg is used for high-quality roasts, in the form of frenched leg of lamb or boneless leg of lamb. The sirloin may be cut into chops for broiling. The leg may also be cut into cubes for a high quality shish kebob.

Tenderization   Age leg 1 week and dry roast slowly.

7.10 SUMMARY

Loin lamb chops, Irish stew, shish kebob, rack of lamb, or any other lamb dish can add variety to the menu. As prices of both domestic and imported lamb continue to rise, lamb is becoming a luxury and thus a very popular and profitable food service item. The information provided in this chapter should be useful in selecting lamb dishes for the menu and in identifying lamb products at the receiving dock.

CASE STUDY   The following food facilities all wish to incorporate at least two lamb items on their menus: a hospital, a university dining program, a resort hotel, and a high-priced family restaurant.

Prepare a list of appropriate luncheon and dinner menu items made with lamb. Use cookbooks or magazines for the menu terminology; then select the appropriate cuts of lamb listed in this chapter to be used for those items. Obtain current market quotations of the price for each item so that you become familiar with the relative cost per pound or cost per portion of lamb cuts.

REFERENCES

1.   Richard D. Biglin, *Lamb Marketing Seminar for American Sheep Producers Council, Inc.*, Published by American Sheep Producers Council, Inc., 200 Clayton Street, Denver, Colorado 80206. (1975).

8 | PORK

Pork production and sales increase when the beef supply becomes short and beef prices rise. The retail consumer shifts demand from beef to pork and poultry when beef prices reach certain levels. The increased pork supply acts as a brake to further increase of beef prices.

Pork produced today yields more meat and less fat, called "lard" in pork, than formerly. Hogs raised for meat production are slaughtered between six and eight months of age. Normal slaughter weights are between 200 and 220 lb, which yields dressed carcasses weighing 140–150 lb.

8.1 SOME FACTS ABOUT PORK

A Popular, Nutritious, and Versatile Meat

Pork is a very popular meat, second only to beef in consumption. Today's leaner pork has been accepted by weight watchers and nutritionists as an excellent source of protein and B-complex vitamins. The many products derived from fresh and cured pork add great variety to the breakfast, lunch, and dinner menu of a food service operation.

PSE Pork

One problem created by the selecting and breeding of a meatier hog has been an increase incidence of **pale, soft,** and **exudative (PSE) pork** muscle. According to Desrosier, "the incidence of this condition averages about 18% and may approach 40 to 50% in the summer."[1]

The major problem with PSE pork is a large increase in shrink loss—moisture loss during storage and cooking. An additional shrink loss of 20 percent above the normal storage and cooking losses can be experienced when processing PSE pork. This decreases the yield of the product and significantly increases the portion cost.

Fresh Versus Cured Pork

Much of the pork produced today is not sold in the fresh state but is further processed into bacon, ham, sausage, and other cured and smoked meat products. (See Chapter 13, Processed Products.) Curing and smoking are methods of preservation that not only increase the refrigerated shelf life of pork but also multiply the ways pork can be used on the menu. Fresh pork has a relatively short shelf life and should be used quickly. When preparing pork roasts and chops, cook to an internal temperature of 160°F—no higher. Many cookbooks and thermometers call for internal temperatures of 170 to 185°F. This is much too high and results in a drier, less tender product.

Trichinosis — Not Really a Problem

Pork is a possible carrier of a nematode (worm) called trichinella spiralis. A person consuming infected pork that has not been cooked properly can become the host for this parasite and may experience nausea or severe muscle cramps. Changes in production and feeding of hogs have resulted in almost total disappearance of trichinosis. The National Live Stock and Meat Board estimates that "less than one-tenth of one percent of today's pork may contain traces of trichina." Even if trichina is present, however, it can easily be destroyed by proper handling. *To guarantee serving trichina-free pork, the food service operator should do one of the following:*

1. Cook pork to an internal temperature of 160°F. (Trichina is actually destroyed at 137°F.)
2. Freeze pork either at 0°F for 30 days or at −20°F for 7 days. Some meat processors are able to sell government-certified trichina-free pork by having a lock-up area in a freezer which is controlled by a USDA inspector.
3. Cure and smoke pork, since trichina is destroyed by curing and smoking.

Given these three alternative methods of guaranteeing safe pork, there really should be no problem in a commercial food service operation. One incident did occur on board a cruise ship in 1975. The food production crew was making fresh pork sausage for breakfast, then grinding hamburger for lunch in the same grinder, without cleaning the grinder after using it for the pork. Diners who consumed the pork sausage were not infected because it was cooked above 137°F. Diners who ordered rare hamburgers at lunch were the victims instead.

8.2 PORK CARCASS BREAKDOWN

Figure 8.1 shows the location of the primal and wholesale cuts on the pork carcass. Table 8.1 and Figure 8.2 show how those wholesale cuts are further broken down and utilized. The remainder of the chapter discusses the major cuts obtained from the pork carcass, how they are identified, and how they can be utilized in a food operation.

Figure 8.1 Pork primal and wholesale cuts.

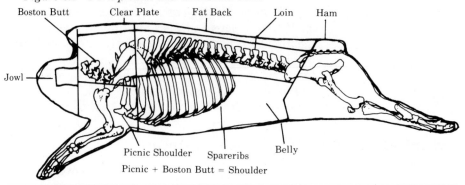

TABLE 8.1 PORK PRIMAL AND WHOLESALE CUTS

Approximate Yields, %

Fresh ham, skinned	21.0
Loins, blade-on	18.0
Boston butt	6.6
Picnic shoulder	8.8
Bacon, square cut	17.3
Spareribs	3.8
Jowl, trimmed	3.0
Feet, tail, neck bones	6.0
Fat back, clear plate, fat trimmings	11.2
Sausage trimmings	4.3
Total	100

Figure 8.2 Utilization of primal and wholesale cuts of pork.

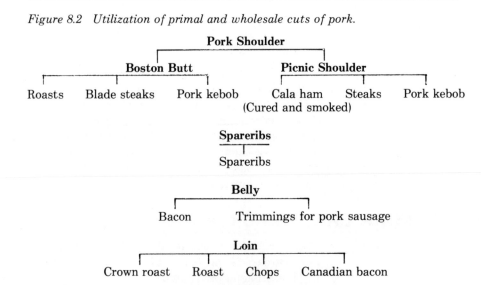

8.3 SHOULDER

Identification
1. Includes the shoulder blade bone and arm bone.
2. Can be broken down into the Boston butt and picnic ham.
3. Approximate weight (range B) 12–16 lb.

Utilization The shoulder can be boned, rolled, and tied as a roast or broken down into its components: the Boston butt and picnic shoulder. It can also be used for pork cubes for sweet and sour pork or pork kebobs. Pork steaks or cutlets may also be cut from the shoulder. These generally require tenderization by cubing by means of a cuber tenderizer.

Boston Butt

Identification
1. Includes the shoulder blade bone.
2. Clear plate is removed, leaving no more than a ¼-in. fat cover.
3. Approximate weight (range B) 8–12 lb.

Utilization The Boston butt can be smoked, cured, or prepared fresh. Generally, the shoulder blade bone is removed, and the butt is rolled and tied for roasting. The butt can also be sliced on the bandsaw for blade steaks or cubed for kebobs. The Boston butt is a popular item for barbecuing whole.

Picnic Shoulder

Top view   *Picnic face where Boston butt is removed*

Identification
1. Includes a portion of the blade bone and the arm bone.
2. Skin and fat are beveled to the equivalent thickness of the fat at the butt end.
3. Resembles the fresh ham.
4. Approximate weight (range B) 6–8 lb.

Utilization The picnic can be cured or prepared fresh. Although sometimes called "picnic ham" or "Cala ham," this cut is less desirable than fresh ham. Similar to the butt, the picnic can be boned, rolled, and tied for roasting or it can be sliced into thin pork steaks for pan frying or braising.

8.4   BELLY

Identification
1. Corresponds to the plate in beef but boneless when spareribs are removed.
2. Contains large percentage of fat.
3. May have rind on or off.

Utilization Bellies are ordinarily cured, smoked, and sold as bacon. In the fresh state they may be ground for pork sausage or ground pork.

8.5   SPARERIBS

Identification

1. Includes portions of the breast bone covered with lean meat and 13 rib bones.
2. Approximate weight (range B) breast on: 3–5 lb.

Utilization Spareribs can be prepared with moist or dry heat. They are popular barbecued or cooked with sauerkraut.

8.6 LOIN

Pork loin

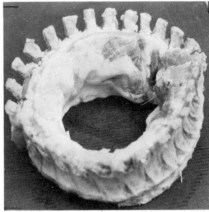

Crown roast

Loin chops

Identification

1. Includes portion of the blade bone, 13 rib bones, and 2 sacral vertebrae.
2. Fat does not exceed more than 1¼ in. in thickness over the major loin muscles except for the hip bone area.
3. Approximate weight (range B) 14–17 lb.

Utilization The loin is generally prepared fresh and dry roasted. It yields the most desirable roasts and chops—center-cut loin chop and rib chop. Cutting the loin end to end produces four types of chops: blade end chop (identify by blade bone), rib chops (identify by rib eye muscle), center-cut loin chops (identify by loin muscle and tenderloin), and loin-end or sirloin chops (identify by portion of the hip bone). These chops can be successfully prepared by pan-frying, broiling, or braising. If the loin is to be prepared for service in the dining room, a crown roast would be appropriate. The loin can also be boned and cured for Canadian bacon.

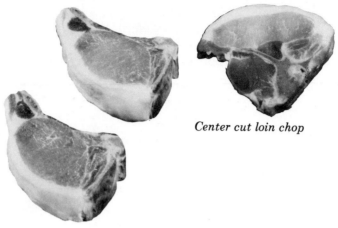

Center cut loin chop

Center cut rib chops

8.7   HAM

Identification
1. Includes the aitch bone, the femur, and a portion of the hip bone.
2. Skin is removed on the outside to approximately 2½ in. in front of the stifle joint (where the femur meets the shank).
3. Approximate weight (range B) 14–17 lb.

Utilization   In the fresh state, the ham produces a very desirable pork roast for commercial use. It can be boned, rolled, and tied for easy carving. Excellent pork cubes or thin pork steaks can also be fabricated from the fresh ham. When cured and smoked as Virginia ham this cut may be baked or cut into ham steaks for broiling or grilling. Most of the top-quality cured and canned hams for slicing are made from this cut.

8.8   SUMMARY

Currently pork is the most economical red meat one can purchase. It is economical, nutritious, and unlike other meats, it can be featured on all three menus: breakfast, lunch, and dinner. This chapter provided information on fresh pork selection and preparation; for information on cured and smoked pork products refer to the Chapter 13 on processed products, which increase pork's versatility as a menu item.

CASE STUDY Because of a shortage of beef and the resulting price rise, a medium-priced restaurant chain is investigating the possibility of adding more pork items to their menu. A major concern to the owner of this chain is the risk of trichinosis and the customers' perceived loss of nutritional value of the meal.

You are hired as a consultant to suggest several breakfast, lunch, and dinner menu items using pork. After selecting these items, in the food composition tables of a current nutrition textbook look up the nutrients supplied by the portion size you have specified. Compare nutrient values of several of these to the nutrient values of other meats. Use this information and information provided in this chapter to dispel the restaurateur's fears.

REFERENCES
1. Desrosier, Tressel, *Fundamentals of Food Freezing.* (Avi, 1977), p. 221.

9 | POULTRY

Poultry is one of the most economical sources of meat. In total consumption poultry is in third place, behind beef and pork. Poultry is very high in nutrition, with a higher percentage of protein and lower percentage of fat than red meat. With the development of **broiler** production, chickens can now be grown to market size in eight to ten weeks, providing a constant supply of young, tender, and flavorful poultry.

Its low price and wide range of products make poultry useful for every type of food service from fast food to gourmet restaurant. To profit from poultry the food service manager must understand the product and merchandise it effectively. Because poultry products are generally considered separately from red meats, this chapter begins with an exploration of the inspection and grading of poultry. It goes on to cover poultry classification, purchasing, fabrication, and merchandising.

9.1 INSPECTION

In 1957, the Poultry Products Inspection Act was passed by the U.S. Congress, assuring consumers in this country that all poultry moving across state lines and imported from foreign sources was federally inspected. Then in 1968, the Wholesome Poultry Products Act, an amendment to the 1957 law, was passed. As a result, poultry inspection is now executed in a manner similar to that required for red meat by the Wholesome Meat Act of 1967. Inspection is administered by the Meat and Poultry Inspection Program of the USDA's Food Safety and Quality Division.

The U.S. Department of Agriculture requires poultry, like red meat, to undergo both ante mortem and post mortem inspection. Prior to slaughtering, the live birds are checked for any visible signs of disease. As with meat, one of three things can happen. If the live birds are free from any visual sign of illness, the inspector passes the birds, allowing them to be slaughtered. If the birds are suspected of carrying a disease, however, the lot is tagged "suspect." The suspect lot is detained to be inspected further before being slaughtered separately from the healthy flock. When the time comes, the suspect birds are closely inspected to determine whether the birds are completely diseased or whether infection is only local. In the latter case the inspector may decide that parts of the bird are indeed safe for human consumption. Any bird judged unfit for human consumption is condemned and disposed of.

Post mortem inspection is the second major procedure. After the slaughter, birds intended for **evisceration**—removal of intestines, spleen, and other organs—are opened to expose the organs and body cavity. The federal inspector reexamines every carcass for any further possible signs of disease. Once again one of three decisions is made. The carcass can be passed, retained, or condemned.

Following inspection, disease-free eviscerated birds are immediately prepared as ready-to-cook poultry: the bird is bled, plucked, and eviscerated, the head and feet cut off, and the giblets are retained. Poultry intended for poultry products is rechecked after shipment to the processing plant, where it is examined from processing to finished product. All processing procedures must be approved by the USDA.

The federal inspection mark is the symbol of wholesomeness. It assures the consumer that the bird was processed under sanitary conditions, contains no harmful chemicals or additives, and was properly packaged and labeled.

USDA Inspection Mark
Found on federally inspected fresh or frozen poultry or processed poultry products.

DRESSED POULTRY

ELIGIBLE FOR FURTHER PROCESSING

In Official Establishments Under USDA Inspection

Plant No. 000.   Lot 000

USDA Dressed Poultry Identification Mark
Found on dressed poultry processed under inspection and passed for distribution.

9.2 GRADING

The federal inspection stamp alone does not give the buyer any indication of the quality of the product being purchased; as with meat, only grading, which is optional, does that. For a fee, individuals, firms, or governmental agencies can obtain the services of a USDA grader. These services are administered by the Grading Branch of the Poultry Division of the Agriculture Marketing Service of the U.S. Department of Agriculture.

Only ready-to-cook poultry and poultry products are graded. Poultry is graded according to three factors: **condition** (wholesomeness), **class** (age and sex), and **quality** (degree of desirability).

Condition: Wholesomeness

All poultry must be inspected prior to grading. The examiner assigns a grade only to poultry carcasses and parts that are fresh. Frozen poultry cannot be graded after it is frozen; it must have been graded when fresh to be identified with the appropriate grade mark.

Ready-to-cook poultry that shows evidence of decomposition is ineligible for grading. Ready-to-cook birds showing protruding pinfeathers, bruises requiring trimming, lungs or sex organs incompletely removed, or the presence of parts of the trachea, or windpipe, vestigial feathers, feathers, or any extraneous material such as blood or fecal material, cannot be graded. They are sent back for further processing.

Class: Age and Sex

Age of poultry is an important determinant of tenderness. It should be taken into consideration when deciding how to use the bird. Young poultry is ideal for roasting, broiling, and frying; older poultry is better suited for braising or stewing. The skin and breastbone indicate the age in poultry; young birds have soft, thin skin and a flexible breastbone, whereas older birds have coarse, thick skin and a rigid and firm breastbone. The species of poultry are divided into classes according to physical characteristics, as shown in the following list:

CLASSES OF POULTRY

Chickens
1. Rock Cornish game hen or Cornish game hen. A young, immature chicken, usually 5 to 6 weeks of age, weighing not more than 2 lb, ready-to-cook weight, which was prepared from a Cornish chicken or the progeny of a Cornish chicken crossed with another breed of chicken. (Cornish is the name of the breed.)
2. Fryer or broiler. Young chicken usually under 13 weeks of age, of either sex, that is tender-meated with soft, pliable, smooth-textured skin and flexible breastbone cartilage.
3. Roaster. Young chicken usually 3 to 5 months of age, of either sex, that is tender-meated with soft, pliable, smooth-textured skin and

breastbone cartilage that is somewhat less flexible than that of a broiler or fryer.

4. Capon. An unsexed male chicken (usually under 8 months of age) that is tender-meated with soft, pliable, smooth-textured skin.

5. Stag. A male chicken usually under 10 months of age with coarse skin, somewhat toughened and darkened flesh, and considerable hardening of the breastbone cartilage. Primarily used for soup.

6. Hen, or stewing chicken or fowl. A mature female chicken usually more than 10 months of age with meat less tender than that of a roaster and with inflexible breastbone.

7. Cock, or old rooster. A mature male chicken with coarse skin, toughened and darkened meat, and hardened breastbone. Used for soup base.

Turkeys

1. Fryer or roaster. A young turkey usually under 16 weeks of age, of either sex, that is tender-meated with soft, pliable, smooth-textured skin and breastbone cartilage that is somewhat less flexible than that in a turkey fryer.

2. Young hen turkey. A young female usually under 8 months of age that is tender-meated with soft, pliable, smooth-textured skin and breastbone cartilage that is somewhat less flexible than that in a turkey fryer.

3. Young tom turkey. A young male usually under 8 months of age that is tender-meated with soft, pliable, smooth-textured skin and breastbone cartilage that is somewhat less flexible than that in a turkey fryer.

4. Hen turkey. A mature hen turkey or old hen turkey, usually over 15 months of age, with toughened flesh and hardened breastbone. It may have coarse or dry skin and patchy areas of surface fat.

5. Tom turkey. A mature tom turkey or old tom turkey, usually over 15 months of age, with coarse skin, toughened flesh, and hardened breastbone.

Ducks

1. Broiler, or fryer, duckling. A young duck usually under 8 weeks of age, of either sex, that is tender-meated and has a soft bill and soft windpipe.

2. Roaster duckling. A young duck usually under 16 weeks of age, of either sex, that is tender-meated and has a bill that is not completely hardened and a windpipe that is easily dented.

3. Mature, or old duck. A duck usually over 6 months of age), of either sex, with toughened flesh, hardened bill, and hardened windpipe.

Geese

1. Young goose. A goose of either sex, tender-meated, with a windpipe that is easily dented.

2. Mature, or old goose. A goose of either sex with toughened flesh and hardened windpipe.

Guineas

1. Young Guinea. Of either sex, tender-meated, with a flexible breastbone cartilage.
2. Mature, or old guinea. Of either sex, with toughened flesh.

Pigeons

1. Squab. A young pigeon of either sex that is especially extra tender-meated.
2. Pigeons. Mature bird of either sex with coarse skin and toughened flesh.

Quality: Degree of Desirability

Several criteria are used to assess quality in poultry: **conformation**, or structure of the bird; **fleshing**, the amount of flesh covering the parts of the carcass; fat covering, quantity of pinfeathers, and number of defects (exposed flesh, cuts, tears, broken bones, skin discolorations, flesh blemishes, bruises, and freezer burns). Based on these criteria , a grade of A, B, or C is assigned to the bird (see Table 9.1.)

The major difference between the three grades is the amount of flesh and fat covering and number of defects. Grade A poultry is well fleshed, has a good fat covering, and is free of defects. Grade A poultry should be purchased if the appearance of the bird on the plate is important, as in chicken Cordon Bleu. Grade B poultry exhibits a few defects and is only moderately fleshed. It should be purchased if the appearance of the bird is not as important, as in chicken cacciatore. Grade C poultry is poorly fleshed and exhibits major defects. Normally, this grade goes to large commercial manufacturers for use in processed chicken products.

U.S. Procurement Graded Poultry

Large institutions generally purchase poultry graded by another system, U.S. Procurement, which places a greater emphasis on yield than on appearance. U.S. Procurement grading is based on sample lots. Each bird in the sample lot is graded but not all lots are examined. There are two grades for ready-to-cook poultry: U.S. Procurement grade I and U.S. Procurement grade II.

U.S. Procurement grade I poultry requires 90 percent or more of the carcasses in the lot to meet the quality standards set for USDA grade A ready-to-cook poultry with a few exceptions, as follows: Fat covering, conformation, skin, and flesh discoloration can be B quality. One or both drumsticks or parts of the wing can be removed if the part is severed at the joint. Skin and flesh can be trimmed to remove defects as long as the yield is not substantially affected and not more than one third of the flesh is exposed on any part. The remaining 10 percent may have only moderate flesh covering but must meet the same requirements as the other grade I carcasses. Carcasses failing to meet the requirements stated for U.S. Procurement grade I poultry can be classified U.S. Procurement grade II if the flesh trim from any one part does not exceed 10

TABLE 9.1 SUMMARY OF SPECIFICATIONS OF QUALITY FOR INDIVIDUAL CARCASSES OF READY-TO-COOK POULTRY AND PARTS THEREFROM (Minimum Requirements and Maximum Defects Permitted)

Factor	A Quality			B Quality			C Quality
Conformation	Normal			Moderate deformities			Abnormal
Breastbone	Slight curve or dent			Moderately dented, curved, or crooked			Seriously curved or crooked
Back	Normal (except slight curve)			Moderately crooked			Seriously crooked
Legs and Wings	Normal			Moderately misshapen			Misshapen
Fleshing	Well fleshed, moderately long, deep, and rounded breast			Moderately fleshed, considering kind, class, and part			Poorly fleshed
Fat Covering	Well covered—especially between heavy feather tracts on breast and considering kind, class, and part			Sufficient fat on breast and legs to prevent distinct appearance of flesh through the skin			Lacking in fat covering over all parts of carcass
Pinfeathers							
Nonprotruding pins and hair	Free			Few scattered			Scattering
Protruding pins	Free			Free			Free

Exposed Flesh (in.)*

Carcass Weight, lb		Breast and Legs	Elsewhere	Part	Breast and Legs‡	Elsewhere†	Part	
Minimum	Maximum							
None	1½	None	¾"	Slight trim on edge	¼"	1½"	Moderate amount of the flesh normally covered	No limit
Over 1½	6	None	1½"		1½"	3"		
Over 6	16	None	2"		2"	4"		
Over 16	None	None	3"		3"	5"		

Discolorations§

		Breast and Legs	Elsewhere	Part	Breast and Legs	Elsewhere	Part	
None	1½	½"	1"	¼"	1"	2"	½"	No limit§
Over 1½	6	1"	2"	¼"	2"	3"	1"	
Over 6	16	1½"	2½"	½"	2½"	4"	1½"	
Over 16	none	2"	3"	½"	3"	5"	1½"	

Factor	A Quality			B Quality			C Quality
Disjointed bones	One			two disjointed and no broken or one disjointed and one nonprotruding broken			No limit
Broken bones	None						No limit
Missing parts	Wing tips and tail[0]			Wing tips, second wing joint, and tail			Wing tips, wings, and tail
				Back area not wider than base of tail and extending half way between base of tail and hip joints			Back area not wider than base of tail extending to area between hip joints
Freezing Defects (When consumer packaged)	Slight darkening over the back and drumsticks. Few small ¼-in. pockmarks for poultry weighing 6 lb or less and ¼-in. pockmarks for poultry weighing more than 6 lb. Occasional small areas showing layer of clear or pinkish ice.			Moderate dried areas not in excess of ½-in. in diameter. May lack brightness. Moderate areas showing layer of clear, pinkish, or reddish colored ice.			Numerous pockmarks and large dried areas

*Total aggregate area of flesh exposed by all cuts and tears and missing skin, not exceeding the area of a circle of the diameters shown.
†A carcass meeting the requirements of A quality for fleshing may be trimmed to remove skin and flesh defects, provided that no more than one-third of the flesh is exposed on any part and the meat yield is not appreciably affected.
‡Flesh bruises and discolorations such as blue back are not permitted on breast and legs of A quality birds. Not more than one-half of total aggregate area of discolorations may be due to flesh bruises or blue back (when permitted), and skin bruises in any combination.
§No limit on size and number of areas of discoloration and flesh bruises if such areas do not render any part of the carcass unfit for food.
[0]In ducks and geese, the parts of the wing beyond the second joint may be removed; if removed at the joint and both wings are so treated.

percent of the meat. Poultry parts may receive a grade II if they do not weigh less than one-half of the weight of the whole carcass and have approximately the same ratio of meat to bone as the whole carcass.

Package Markings For Grading

A variety of marking emblems are used for graded poultry:

USDA Grade Stamp
Assures consumer that the product was graded by a U.S. government grader according to the specified standards of quality.

International Trade Development Emblem:
Excellent Quality
Found on poultry products intended for export that conform to quality, freezing, and packaging specifications developed by the U.S. poultry industry's International Trade Development board and examined by an official representative of the U.S. Department of Agriculture. Product must be better or equal to USDA Grade A to be grade Excellent.

International Trade Development Emblem:
Approved Quality
Same criteria as poultry graded Excellent, but product must be better than or equal to USDA Grade B.

Wing Tags and Clips
Tag includes inspection mark and/or grade mark, and plant number or firm address. Tags showing grade mark must also show class of poultry (such as frying chicken) or include a qualifying term (young, mature, or old).

9.3 PURCHASING POULTRY

Advantages of Ready-to-Cook Poultry

In the United States today almost all poultry is sold in ready-to-cook form rather than live. The bird is plucked, bled, eviscerated, head and feet are removed, and giblets are separately wrapped. Birds are processed this way for three major reasons:

1. Sanitation. The carcass keeps much longer with viscera removed.
2. Economics. Why ship 20–36 percent waste? (See Table 9.2, Weight Losses in Dressing Poultry.)
3. Convenience. Someone else plucks and eviscerates the poultry.

The exception to ready-to-cook form might be a game bird like a pheasant, which is used uneviscerated, displayed with its plumage on as a centerpiece on a gala buffet.

Besides whole birds various poultry parts, such as wings, breast, and leg, may be purchased in the ready-to-cook form. Prepared and fully cooked poultry items are also available, like Rock Cornish game hen with wild rice stuffing. Fully cooked turkey breasts and cured and smoked poultry products are also very useful.

TABLE 9.2 WEIGHT LOSSES IN DRESSING POULTRY

Class of Poultry	Weight Loss from Live to Ready-to-Cook, %
Broiler or fryer	35–38
Roaster	31–35
Stewing hens	31–36
Hen turkey	22–24
Tom turkey	19–22
Ducks	28–31
Geese	25–30

Source: Dodge and Stadelman, 1959.[1]

One pound of ready-to-cook chicken yields 8 ounces of cooked meat. A 2-pound bone-in fryer is usually sufficient for two servings. Table 9.3 shows what percentage of the dressed bird each part constitutes.

TABLE 9.3 PERCENTAGE OF EACH PART IN READY-TO-COOK POULTRY

Poultry Part	Percentage of the Whole Bird
Breast	22.4
Legs	15.0
Thighs	15.9
Wings	12.6
Back	17.3
Neck	8.2
Heart	.9
Gizzard	3.7
Liver	3.6

Source: Percentage from Dodge and Stadelman, 1959.[1]

Purchasing in Bulk

Poultry is still relatively inexpensive, and there are certain seasons when prices are especially low. Depending on freezer space and funds available, purchasing bulk quantities of poultry may prove to be a worthwhile investment. Broiler-fryer chicken supplies are most abundant from June through October. Prices tend to be higher due to the increased consumer demand during those months, however. Prices are generally lower in the winter months, from November to May.

Turkeys are slaughtered in September through December. The selling prices for frozen turkeys usually decrease as inventories in the freezers increase. Bulk buying of frozen turkeys during those months may yield substantial savings to a food service operation. A note of caution: it is important to avoid purchasing turkeys left over from preceding year's crop if one intends to store them for a long time.

March through October is the best time to purchase ducks.

Writing Specifications

Specifying to the purveyor exactly what is wanted when purchasing poultry is essential for maintaining quality standards and keeping costs down. Writing accurate, detailed descriptions specifying what is needed assures the food operator that the products being ordered will meet the needs of his or her operation. There are six characteristics to include when writing a specification for poultry:

1   Kind. Refers to the species.
2. Type. Fresh or frozen.
3. Class. Physical characteristics depending on age and sex.
4. Size. The weight.
5. Style. The way it was processed.
6. Grade. Quality designation.

One might write the following specification for turkey:

Turkey Specification

Kind:	Turkey
Type:	Fresh chilled, 40°
Class:	Tom
Size:	22–24 lb.
Style:	Ready-to-cook
Grade:	A

The different classes of poultry are available in different weight ranges. Table 9.4 is a guide for sizing poultry when writing specifications for purchasing.

TABLE 9.4 POULTRY SIZES

Class of Poultry	Weight Range
Broilers and fryers	2–2¼ lb, 2¼–2½ lb, 2½–2¾ lb, 2¾–3 lb, 3–3¼ lb, 3¼–3½ lb
Cornish game hen	1–1¼ lb, 1¼–1½ lb, 1½–1¾ lb, 1¾ lb–2 lb
Ducks	4–4½ lb, 4½–5 lb, 5–5½ lb
Young tom or hen turkey	10–12 lb, 12–14 lb, 14–16 lb, 16–17 lb, 17–18 lb, 18–20 lb, 20–24 lb, 25–30 lb

9.4 PROPER HANDLING OF POULTRY

Poultry purchased fresh should not be allowed to stand in its own juices. Poultry iced or packaged in plastic wrap and stored at 28°F will keep for approximately ten days. At a temperature of 34–36°F raw poultry can be stored for no more than six days. Whole birds are generally packaged with the giblets—liver, gizzard, neck, and heart—in the body cavity. These should be removed as soon as possible because they tend to spoil faster than the rest of the bird. If the giblets have begun to decay, remove and discard them and rinse the entire bird thoroughly with cold water. If the odor does not dissipate after this procedure, the entire carcass should be discarded. Frozen chickens can be stored from eight to twelve months at 0°F if packaged so that they are air-tight. When chickens have been frozen, a reddish discoloration occurs in the meat adjacent to the bone. This color may remain even after the bird is thoroughly cooked, but it is harmless.

It is important to remember that raw poultry can contaminate the cutting utensils and surfaces that come in contact with it prior to cooking. It is essential to separate raw poultry from cooked poultry and to sanitize any cutting utensil that has come in contact with the raw product. Once cooked, chicken may be frozen; otherwise, it should be used within five days.

9.5 POULTRY FOR PROFITABILITY

Poultry is ideally versatile. Its low cost makes it an extremely profitable menu item. Unfortunately, many food service operations do not capitalize on the profitability of poultry items even though as much as 45 lb of poultry products are consumed per person each year in the United States. The major argument against serving poultry used by most food service operations is that their customers do not want to bother with the bones when they are dining out. Another argument is that customers can prepare bone-in poultry items easily at home, moreover, and so do not order them in restaurants. One way to sell more poultry, therefore, is to bone it for fancier boneless chicken entrees. Labor costs should be minimal since an experienced person can bone a chicken in a matter of minutes. Should labor be a problem, the boneless product can be purchased at slightly higher prices from the supplier.

Chicken is especially suited for boning. A 3½-lb broiler is ideal for chicken Cordon Bleu or chicken Kiev; a 2-lb chicken is best suited for a stuffed half boneless squab. In chicken Cordon Bleu or Kiev, only the breast and the first section of the wing is used. One 3½-lb broiler can thus be used for two breast entrees, while the leg and thigh can be boned and stuffed for a boneless stuffed chicken entree perfect for lunch. In preparing a boneless stuffed squab, both the breast and thigh with leg are used. Consequently, a 2-lb broiler can yield two boneless entrees at rather low cost. Duck is more salable when only the leg, thigh, and wing bone remain in the cooked portion. The backbone and breastbone are easily removed after roasting the duck. This makes duckling much more enjoyable for the guest to eat. The following section describes step by step the proper procedure for boning a chicken.

Boning A Chicken

STEP 1 Place the chicken, breast side up, on cutting board with its legs facing you. With one hand, grasp carcass at the base of the wings. With the other hand, insert a sharp knife at the top of the breastbone. Begin cutting along the side of the breastbone, keeping the knife close to the bone. Continue cutting along the carcass, pulling the meat back with fingers until reaching the base of the rib cage. Apply pressure at the joint where the thigh bone meets the backbone until it pops. Cut between this joint, separating the thigh from the backbone. Next locate the joint where the wing bone meets the rib cage; cut between the joint, thus completely separating one side from the carcass. Repeat this procedure on the other side. The next step will depend upon the intended use of the product.

STEP 2—2-LB. BROILER FOR BONELESS SQUAB Prepare as in step 1. Remove wing joint, which can be later used for hors d'oeuvres or stock. Next bone the thigh and leg bone. Insert a knife along the bone on the inside seam. Grasp the bone while cutting the meat away until left with a completely boneless cut. Place the boneless chicken, skin side down, on the table or counter. Fill the center with a suitable stuffing; then roll the sides of the chicken to the center, making sure the stuffing remains inside. Shape to desired form and bake, seam side down, until the internal temperature reaches 165°F.

STEP 3—3½ LB BROILER FOR CHICKEN CORDON BLEU OR KIEV Prepare as in step 1. Next separate the leg and thigh from the breast at the natural seam. Remove the top two joints of the wing, leaving one joint on the breast. Pull out the filet. Place breast, skin side down, on the table. Separate the two breast muscles by hand, making the breast as wide as possible. A mallet may be used to flatten the breast a little more. For chicken Cordon Bleu, place a small piece of Swiss cheese in the center of the breast; then put a piece of ham or Canadian bacon on top of the cheese. Top the ham with the flattened filet. Roll the breast from side to side to seal in the ingredients.

Dip the breast first in flour, then in an egg-and-milk batter, and then in bread crumbs. Brown the breaded breast in hot oil in a deep fryer basket, with the wings resting against the side of the basket. The protein will coagulate in the wing joint, keeping it in the upright position. After it is browned, the breast can be baked until the internal temperature reaches 165°F or the chicken can be frozen for later use. Chicken Kiev is prepared with an herb butter stuffing instead of ham and cheese.

If a market exists for chicken legs and thighs, the leg and thigh can be boned and stuffed with a suitable stuffing as for boneless half squab or the thigh and leg can be separated (bones intact for each cut) and prepared as fried chicken. If no market exists for these by-products, just the breast with wing on can be purchased.

Boneless chicken entrees can be extremely profitable, competing in menu price with beef, lamb, and veal dishes that are more expensive to prepare. At current prices, it would cost approximately $1.10 to prepare a chicken Cordon Bleu. It is not common to see this item on a restaurant menu for $7.50 or more. Not many items can yield as much profit and still please the customer.

Turkeys

Turkeys deserve special mention. Approximately 75 percent of all turkeys are sold frozen. Traditionally, they are purchased whole and roasted. They are also available as bone-in or boneless turkey breasts, fully cooked turkey breasts, turkey rolls, and soy-extended turkey rolls and turkey "hams." Turkey has great potential for use in the food service industry. With the rising cost of red meat, food service operations would do well to introduce lower priced entrees on their menus. Turkey is ideal for this purpose, being of low cost per pound compared with red meat, yet equally versatile. Because of turkey's natural tenderness, turkey breasts can be sliced as cutlets and prepared like veal—as turkey marsala, scaloppine, française, and so on. Four to six ounces of boneless turkey (or chicken) is a sufficient portion, and it can be lower priced on the menu and still be more profitable to the food operation than red meat. New and exciting entrees can be created for the guest to experience. The average customer would not prepare a turkey or chicken Cordon Bleu or a turkey cutlet française at home. Such additions to the menu have first-class status and are disassociated from the run-of-the-mill fried chicken or baked half broiler. Turkey breasts, moreover, have the highest cooked protein content compared to other meat products, as shown by Table 9.5.

TABLE 9.5 PROTEIN AND FAT CONTENT OF POULTRY AND MEAT

	Protein, %	Fat, %
Turkey breast	33–35	6–8
Turkey leg and thigh	30	11
Chicken breast	31.5	1.3
Chicken leg	25	7.3
Beef	21–27	13–32
Pork	24	26–33
Lamb	24	19–30
Veal	28	11

Roast Turkey

As most turkeys are sold frozen, food operations must be prepared to thaw the product under refrigeration from two to four days. Turkeys can be cooked from the frozen state, but the cooking takes much longer. Start at 200°F until the bird is thawed and then raise the temperature to 325° until it is done. The turkey is ready to be removed from the oven when the internal temperature reaches 165°F.

Since the breast cooks faster than the legs and thighs, the white meat can be cooked separately if it is separated from the dark meat. Cut the skin between the breast and thigh, then grasp the tip of the breastbone with one hand and the tail with the other, pull down, and the breast will separate from the legs and thighs at the backbone. Cook all breasts in one oven, all legs and thighs in another. This reduces cooking time and allows for easier carving. The breast meat may be removed from the bone and portioned on a slicing machine.

Allowing the roasted turkey to stand prior to carving will facilitate carving and prevent excess bleeding of juices. Turkeys to be used for sandwiches or cold plates may be cooked a few days in advance and refrigerated. Fully cooked turkey breasts can also be purchased as an alternative to roasting turkeys or as a back-up for busy days.

In terms of yield, larger turkeys are a better buy, as indicated by Table 9.6.

TABLE 9.6 TURKEY YIELD

Raw, lb	Cooked, lb	Number of 5-oz portions
10–12	4½	12
12–14	5	15
14–16	6	18
16–17	6⅔	20
17–18	7⅓	23
18–20	8	25
20–24	10½	33
25–30	13½	40

Ducks

Most ducks are purchased frozen. The weight range is usually 4½ to 5 pounds. Unlike chicken broilers or fryers, for which 1 to 1½ pounds is sufficient for a serving, a full half duckling is generally required for one serving.

Ducks contain a high percentage of fat under the skin. Much of this fat must be rendered during the roasting process. The finished duck should have a crisp exterior skin with a relatively fat-free flesh that is still juicy and tender. To accomplish this, the ducks should be roasted on a rack over an oven pan. The rack keeps the duck from frying in its own grease. The pan should be high enough to keep the grease from spilling over and causing a fire or injury to the person removing it from the oven.

It may require 1 to 1½ hours to roast a duck. For this reason, many restaurants par-cook ducks and finish the cooking when the order is placed. (Most menus inform the guest of a half hour wait for roast duck.) Although this practice is not as desirable as preparing duck from the fresh state, the time required as well as the cost of preparation make this the only feasible method at this time. After the duck is cooked, the rib cage and backbone should be removed, leaving bones only in the leg, thigh, and wing. This service will be appreciated by the customer and will increase the salability of the item.

Table 9.7 illustrates the average yield of the different parts of the duck.

TABLE 9.7   PERCENTAGE YIELD OF A 7½-WEEK PEKIN MALE DUCK*

Breast	29.7	Neck	5.4
Leg and thigh	23.4	Heart	.9
Wings	10.6	Gizzard	4.3
Back	23	Liver	2.7

Source: Percentages from Clements and Winter, 1956.
*Swanson, Carlson, and Fry, "Factors Affecting Poultry Meat Yields," Agricultural Experimental Station, University of Minnesota, Bulletin #476.

9.6   SUMMARY

Poultry producers and processors are striving to meet the needs of the food service industry in product type, size, quality, and quantity. New, more convenient products are introduced constantly. One supplier is currently offering portion-controlled turkey cutlets. Other products such as turkey "ham" and precooked duckling are on the market.

The food service manager must be knowledgeable about poultry in order to be receptive to new products and to profit by merchandising them along with other economical poultry products. Entrees such as chicken Cordon Bleu and chicken Kiev have elevated chicken to a higher place on the menu and higher profits at the cash register.

CASE STUDY

1. M. M. Catering, your employer, offers the following poultry items for banquets and buffets:

- Chicken Salad
- Chicken Cordon Bleu
- Chicken Imperial
- Chicken à la King
- Roast Turkey and Dressing
- Sliced Cold Turkey Platter (White Meat)
- Roast Duckling
- Game Hen with Wild Rice

You are asked to write purchasing specifications for the poultry products to be ordered to prepare the items listed. Discuss your decisions regarding kind, class, size, grade, and style (market form) of the products you "spec."

2. A friend of yours has opened a new restaurant that features several poultry entrees. Having studied the market trends she knew that broiler-fryers were cheaper in April and so she bought 1,000 lb and froze them.

It is now July and she is having all sorts of problems. First, she noticed that a case of birds, gizzards and all, that were thawed in the refrigerator had an objectionable odor after five days. Next, she noticed that another case of birds, when cooked, retained a reddish discoloration in the meat adjacent to the bone. Third, she noticed that several of the kitchen help became ill with vomiting, diarrhea, and fever one day after they had been boning chickens for chicken Cordon Bleu. She rememberd they had eaten chef's salad about the same time they were boning the birds. They had prepared their own salads.

Your friend asks your advice. You have checked the freezer and refrigerator and know the temperatures are being properly maintained, so why all the problems? Should your friend be overly concerned in each case? What are your recommendations?

REFERENCES

1. J.W. Dodge, W.J. Stademan, "Does Meat Yield Vary Between Crosses?" *Poultry Processing*, (November 1959), vol. 65:8–9.
2. P. Clements, A.R. Winter, "Duck Shrinkage From Farm To Table and Cooked Edible Meat Parts," *American Egg and Poultry Review* (September 1956), vol. 18:44.

10 | FISH AND SHELLFISH

In 1978, fish consumption in the United States amounted to only twelve pounds per person. The huge difference between fish consumption and consumption of red meat and poultry exists for several reasons. The first is the influence of environment and location: people near the seashore tend to consume much more fish and shellfish than the inland population. Second, a lack of standardization in fish products has made it difficult for the consumer to judge the product prior to purchasing it. Moreover, unsuitable cooking methods often create an undesirable culinary result; methods and timetables used for meat products cannot be successfully transferred to fish preparation. Finally, the most popular items—lobster, shrimp, crabmeat—are very high in price and are in limited supply.

Because the typical consumer does not have the knowledge to select the right type of seafood for particular uses or the skill to handle and cook seafood so that it will be flavorful, fish and shellfish are eliminated from the dinner table in a great many homes. A food service operator can capitalize on the consumers' lack of expertise by purchasing, preparing, and presenting very desirable seafood appetizers and entrees.

The Future of Fish Production

With the advent of the 200-mile limit on foreign fishing in domestic offshore waters and with conservation, it is likely that the supply of fish and shellfish will improve and their prices remain relatively stable. (Conservation is accomplished by establishing harvesting seasons or setting catch limits.) Aquaculture, or fish farming, is another reason for optimism about the seafood supply. One million pounds of trout or catfish can be produced on one acre of land under proper aquaculture conditions.

There are over 50,000 varieties of fish in the sea. At present, only 240 species are marketed, but as the prices of other meats increase, more and more varieties will be utilized.

Consumer Awareness and Nutritional Pluses Make Seafood a Good Menu Item

Many weight-conscious consumers are switching to seafood as a major source of protein. Fish and shellfish supply high-quality protein that is

129

very easily digested (85–95 percent digestibility). Fish is also high in the B-complex vitamins, vitamin B_6, B_{12}, and niacin, and gives a good supply of phosphorus, potassium, and iron. The amount of fat (oil) present in fish varies according to the species, but it is generally well below 15 percent. Nutritionally, fish is low in calories, low in sodium, high in protein, and high in polyunsaturated fats, making it an excellent addition to any diet.

The food service operator attempting to cash in on the benefits of seafood—its availability, variety, high nutrition, and low cost—must have a good working knowledge of fish and shellfish. In this chapter many seafood products will be discussed. Market forms, quality characteristics, availability, preparation techniques, and cooking methods will be given for the major species of fish and shellfish.

10.1 INSPECTION AND GRADING

Inspection by U.S. Department of Commerce

Unlike meat inspection, which is mandatory, fish inspection is voluntary. The U.S. Department of Commerce (USDC) provides inspection services to processors who wish to display the official USDC grade or inspection shield on their product labels. Firms not using the USDC inspection generally set their own standards to ensure quality.

FEDERAL INSPECTION MARK
- Ensures product is pure, wholesome, and safe
- Ensures product is properly labeled
- Ensures product was packaged under sanitary conditions
  Note: At this point the product has not been graded.

The following frozen products may bear the USDC federal inspection mark:
- Raw breaded and precooked breaded fish fillets, portions, and sticks
- Fresh or frozen whole or dressed fish
- Raw breaded shrimp
- Whole cooked crabs, crabmeat, legs, and claws
- Fried fish cakes
- Mixed seafood cakes
- Fried clams and clam cake dinners
- Raw breaded and precooked scallop dinner
- Raw and raw breaded scallops
- Raw fish steaks
- Raw peeled and deveined shrimp
- Fish and shellfish in sauce

Grading by National Marine Fisheries

Fish may also be assigned a grade by the National Marine Fisheries Services. Grading is paid for by the processor, with fish receiving grades as follows:

- Grade A. Best quality, uniform size, excellent condition, optimum flavor, few or no defects.
- Grade B. Good quality but with less uniformity of size and more defects than grade A.
- Grade C. Fairly good quality but with little uniformity and more defects than grades A and B. Nutritional value is generally equal to that of grades A and B.

U.S. GRADE SHIELD Ensures product was graded according to government standards

Frozen products that may bear the U.S. grade shield include

- Fish fillets and fillet blocks
- Raw fish portions and fish steaks
- Raw breaded and precooked fish portions and fish sticks
- Raw headless and raw breaded shrimp
- Raw and precooked breaded scallops
- Raw headless whiting

How to Judge Quality of Fish

Because both inspection and grading are currently optional and affect only a small percentage of the seafood marketed, it is imperative that the food service operator understand how to judge wholesomeness and quality in seafood products. Tables 10.1–10.3 list the desirable and undesirable characteristics to look for when purchasing and receiving fresh and frozen fish.

Fresh versus Frozen

With modern freezing techniques, high-quality fish products are available almost anywhere. Fast-frozen fish can be as good and are sometimes better than fresh fish. Today freezing is done onboard ship or immediately after the ship reaches the processing plant. Fresh fish, on the other hand, may be held onboard ship from one to eight days, then shipped to market. A wholesaler then picks up the fish and may keep it a few more days before he sells it. Some so-called fresh fish may be one week old by the time the food service operator receives it.

In locations where fish is available directly from day boats or fishermen, there is no substitute for the fresh product. Inland, however, or for the variety of species not found locally, properly handled frozen fish are delicious.

TABLE 10.1 CHARACTERISTICS OF FRESH WHOLE FISH

Desirable	Undesirable
Skin	**Skin**
Bright	Dull
Shiny	Cracked or dry (if improperly iced)
Natural slime is transparent or white.	Yellowish or milky slime (possibly slime was washed off to cover up poor condition.)
Scales	**Scales**
Cling tightly to the skin	Loose
	Many missing
Gills	**Gills**
Bright red in color	Light pink, gray, brownish-green in color
Free of slime	Foul odor
Eyes	**Eyes**
Bright	Dull or cloudy
Transparent or clear	Sunken (possibly absent in very rotten fish)
Protruding	
Flesh	**Flesh**
Firm	Soft
Elastic to touch	Separates from backbone
Bright	Leaves a slight imprint when pressed with finger
Leaves no imprint when pressed with finger	Dull
Sticks firmly to bone	
Odor	**Odor**
Mild	Strong, disagreeable fish smell
Fresh sea water	
Abdomen and belly	**Abdomen and belly**
Clean	Discolored (browning)
Agreeable odor	Foul odor
Firm	Soft
Blood	
Red color	Dark color
Normal consistency	Thin consistency

TABLE 10.2 CHARACTERISTICS OF FRESH FISH FILLETS

Desirable	Undesirable
Flesh	**Flesh**
Firm	Soft
Elastic to the touch	Not elastic
Muscles do not separate	Muscles separate readily
Leaves no imprint when pressed firmly	Mushy texture, which leaves a distinct impression when pressed firmly
Odor	**Odor**
Fresh	Strong off-odor present
Mild	

Table 10.2 (Cont.)

Desirable	Undesirable
Feel	Feel
Smooth	Sticky
Clean feel to the touch	Slimy
	Gummy feel to the touch
Color	Color
Bright color characteristic of that species	Dull
	Darkening around the edges
Good quality	Poor quality
Completely free from bones	Sections of the back or pin bone remain in the fillet.

TABLE 10.3   CHARACTERISTICS OF FROZEN WHOLE FISH

Desirable	Undesirable
Flesh	Flesh
Solidly frozen	Partially thawed or thawed
Firm	Soft
Glossy	Evidence of freezer burn or desiccation
No discoloration	Yellowing
Eyes	Eyes
Bright	Dull
Full	Sunken
Gills	Gills
Bright red	Browning
Odor	Odor
Free from odor	Foul odor
Unwrapped	Unwrapped
Fish completely coated with glaze	Fish partially coated with glaze or no glaze
Wrapped	Wrapped
Little or no air space between fish and moisture-proof wrapping	Frost between fish and wrapping
	Wrapping loose

10.2 CATEGORIES OF FISH

Fish will be discussed in three groups: 1) freshwater fish, from lakes and rivers, 2) ocean fish, and 3) shellfish, divided further into the categories of a) true shellfish (mollusks), consisting of clams, oysters, snails, abalone, and squid and b) crustaceans such as lobsters, crabs, and shrimp.

Each species is covered separately, with detailed information as to where the species is generally caught, when it is most economically purchased, average weight, best cooking method, and use. Additional information on the methods of cleaning, dressing, and filleting whole fish as well as shucking clams and oysters is provided.

Market Forms of Freshwater and Ocean Fish

WHOLE
Fish as they come from the ocean or lake.
Edible portion: 43–47%

DRAWN
Eviscerated.
Edible portion: 46–50%

DRESSED
Scaled and eviscerated (head, fins, and tail are usually removed).
Edible portion: 65–69%

STEAKS
Cross-section slice of fish cut usually ¾–1 in. thick (cut across grain).
Edible portion: 84–88%

FILLETS:
Boneless meaty side of fish.
Edible portion: 100%

BUTTERFLY:
Fillet cut in half except for small piece of flesh holding the two sides together.

PORTION:

Cut from frozen fish blocks and breaded.

Raw breaded: usually ⅜ in. thick, weighing greater than 1½ oz and must contain not less than 75% fish flesh.

Fried breaded: similar to raw breaded but must contain not less than 65% fish flesh.

STICKS:

Cut from frozen fish blocks, breaded, and usually 1 in. x 3 in. in size.

CHUNKS:

Cut from cross sections of dressed fish with only a piece of the backbone remaining.

Table 10.4   SPECIES OF FISH GENERALLY CLASSIFIED AS FATTY

Barracuda	Grayling	Mullet	Smelt
Bass, Striped	Grouper	Pompano	Snappers
Bluefish	Grunt	Porgies	Sturgeon
Bonito	Halibut	Salmon	Swordfish
Carp	Herring	Sardines	Tuna
Croaker	Kingfish	Shad	Whale
Eel	Mackerel	Shark	Whitefish

TABLE 10.5   SPECIES OF FISH GENERALLY CLASSIFIED AS LEAN

Bass, Channel	Flounder	Muskey	Sheepshead
Blackfish	Fluke	Perch	Skate
Bluegill	Haddock	Pickerel	Sucker
Catfish	Hake	Pike	Sunfish
Cod	Lemon Sole	Pollock	Weakfish
Drumfish	Marlin	Scrod	Whiting

The following pages show the major species and give information as to origin, size, market forms, and cooking methods. Where possible the "best buy months" are indicated. These are months when the particular species is most plentiful in the marketplace.

10.3 FRESHWATER FISH

Rainbow Trout

Origin: Inland rivers, imported (Denmark, Japan), freshwater streams throughout the United States, aquaculturally produced in the United States.
Fat or lean: Fat
Weight: ⅓–2 lb.
Market form: Drawn, dressed, boned, boned and breaded, fresh, frozen
Cooking method: Pan fry, deep fry, broil, stuff and bake, poach.

Lake trout

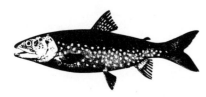

Origin: Great Lakes and imported
Fat or lean: Fat
Weight: 1½–10 lb (usually 4–5 lb)
Market form: Drawn, dressed, fillets, steaks, fresh, frozen
Cooking method: Broil, stuff and bake, sauté

Channel Catfish

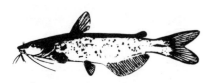

Origin: Great Lakes, other U.S. lakes, inland rivers, ponds, creeks, fish farmed in the southeastern U.S.
Fat or lean: Lean
Weight: ½–2½ lb
Market form: Whole, dressed, fillets, fresh, frozen
Cooking method: Bake, broil, grill, smoke, sauté, barbecue

Yellow Perch

Origin: Great Lakes, other U.S. lakes, inland rivers, Canada
Fat or lean: Lean
Weight: ½–2 lb
Market forms: Whole, drawn, pan dressed, fillets, fresh, frozen
Cooking method: Deep fry, pan fry, bake

10.4   OCEAN FISH

Haddock

Origin: New England, Canada, Iceland, Norway, England
Fat or lean: Lean
Weight: 1½–7 lb
1½–2½ lb for scrod (small haddock)
Market form: Whole, drawn, fillets, fresh, frozen, salted, smoked
Cooking method: Broil, bake, sauté
Best buy month: All except January

Cod

Origin: North Atlantic, North Pacific, New England, Middle Atlantic, Pacific Coast, Iceland, England, Norway, Germany, Denmark, Canada
Fat or lean: Lean
Weight: 3–20 lb for cod, 1½–2½ lb for scrod (small cod)
Market form: Drawn, dressed, steaks, fillets, breaded and precooked sticks and portions, fresh, frozen, salted
Cooking method: Bake, deep fry, sauté, broil
Best buy months: March through May

Flounder

Blackback, or winter

Varieties of Flounder

Species	Weight, lb	Origin
Dab	¾–2½	New England
Gray sole	¾–4	New England, imported
Blackback or winter	¾–2	New England, Middle Atlantic, imported
Lemon sole	¾–4	New England, imported
Fluke or summer	2–12	New England, Middle Atlantic
Dover sole	½–2½	Pacific Coast

Fat or lean: Lean
Market form: Whole, fillets, breaded portions, stuffed, fresh, frozen
Cooking method: Broil, bake, sauté, deep fry

Fluke, or summer

Halibut

Origin: Pacific Coast, Alaska, New England
Fat or lean: Lean
Weight: 5–75 lb
Market form: Drawn, dressed, steaks, fillets, fresh, frozen, breaded portions, smoked
Cooking method: Bake, broil, sauté
Best buy months: June through November

Ocean Perch

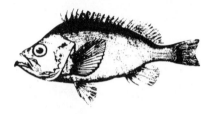

Origin: New England, Iceland, Germany, England, Norway, Canada
Fat or lean: Lean
Weight: ½–2 lb
Market form: Whole, fillets, fresh, frozen, breaded raw or cooked fillets and portions
Cooking method: Pan fry, bake, broil, sauté
Best buy months: May through September

Pollock

Origin: New England, imported
Fat or lean: Lean
Weight: 1½–12 lb
Market form: Drawn, dressed, steaks, fillets, fresh, frozen, breaded raw or cooked sticks and portions, salted, smoked
Cooking method: Bake, broil, pan fry, steam, poach
Best buy months: January, February, August through December

Red Snapper

Origin: Gulf, Middle Atlantic
Fat or lean: Lean
Weight: 2–20 lb
Market form: Drawn, dressed, steaks, fillets, fresh, frozen
Cooking method: Poach, bake, broil

Salmon

Varieties	Weight, lb	Origin
Chinook	5–30	Pacific Coast, Alaska, imported
Sockeye	3–5	Pacific Coast, Alaska
Pink	4–10	Pacific Coast, Alaska, imported
Coho (Silver)	5–18	Pacific Coast, Alaska, imported
Chum	5–11	Pacific Coast, Alaska, imported

Fat or lean: Fat
Market form: Dressed, steaks, fillets, fresh, frozen, smoked, canned
Cooking method: Poach, bake, broil
Best buy months: June, July

Swordfish

Origin: New England, Middle Atlantic, Pacific Coast, Chili, Peru, Japan
Fat or lean: Lean
Weight: 50–4,000 lb
Market form: Dressed, steaks. fresh, frozen
Cooking method: Grill, broil, bake

Striped Bass

Species: Sea Bass
Origin: North America, Chesapeake Bay, Albemarle Sound
Fat or lean: Fat
Weight: 2–25 lb or more
Market form: Whole, drawn, steaks, fillets, frozen, fresh
Cooking method: Bake, pan fry, broil

Weakfish

Species: Sea Trout
Origin: Middle and South Atlantic
Fat or lean: Lean
Weight: 1–10 lb
Market form: Whole, drawn, dressed, fillets
Cooking method: Sauté, bake, broil

Whiting

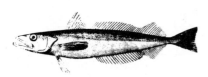

Origin: New England, Middle Atlantic, North Atlantic, England
Fat or lean: Lean
Weight: ½–4 lb
Market form: Whole, drawn, dressed, fillets
Cooking method: Bake, broil, pan fry, poach, deep fry

Cleaning and Dressing Freshwater and Ocean Fish

Step 1. Scaling
Rest the fish against a cutting board. With one hand, grasp the tail of the fish securely. With the other hand, use a scaler or hold a boning knife vertically against the fish and scrape off scales from tail to head.

Step 2. Cleaning
Insert tip of knife at the vent (the anal opening). Cut belly open from vent to head. Remove entrails. Cut around pelvic fins and remove them.

Step 3. Removing the Head
Insert the knife just behind the pectoral fin. Cut through flesh and bone on both sides to remove the head. If necessary, place the head of fish over the edge of the table and apply enough force to break backbone and remove head. If desired, cut off the tail.

Step 4. Removing the Fins
Cut flesh along both sides of the dorsal (back) fin. Pull the fin toward the head making sure to remove root bones. Remove all other fins the same way.

Note: Cutting fins off does not remove root bones. They must be pulled out.

Cutting Fish Steaks

Method
Place the dressed fish on a cutting board. Cut the fish crosswise into steaks about 1 in. thick.

Filleting

Step1
Place the fish, side down, on a cutting board. Next, place one hand on top of the fish. With the other hand, insert a knife and cut along the back of fish from tail to head. Continue cutting down to the backbone.

Step 2
Place the knife flat against backbone. With the other hand, grasp the flesh. Cut along the backbone, head to tail, pulling the flesh back as you cut. Remove the flesh in one piece.

Step 3
Remove the fillet in one piece. Repeat on the other side. Check for bones by running fingers along the fillet.

Step 4
Place fillet, skin side down, on the cutting board. Hold the end of tail firmly with fingertips. At tail end, insert knife and cut through the flesh to the skin. Next, turn the knife flat against skin, with the sharp edge facing head. Begin cutting the flesh from the skin while pulling the skin away from the flesh with the fingers.

Note: If the fillet is to be skinned there is no need to scale the fish prior to filleting. An alternate method of removing skin from fillet is the following: At time of cooking, place the fillet, skin side up, on a broiler pan. Broil for about 30 seconds. With pliers, grasp the skin at the tail and pull toward the front of the fillet. (The heat breaks the bond between the skin and flesh.)

10.5 SHELLFISH

Shellfish are available in the following market forms:

•	Live, in shell.	Alive when sold (lobster, clams, oysters, crab).
•	Cooked, in shell	Species cooked in shell and then chilled or frozen before being sold (crab, lobster).
•	Shucked.	Meat is removed from shell (clams, oysters, scallops). Proportion edible: 100%.
•	Headless	Only tail remains (shrimp, spiny lobster tail). Proportion edible: 50–85%.
•	Cooked meat.	Cooked flesh only (shrimp, crab, lobster meat). Proportion edible: 100%.

Clams

Soft clams

Hard clams
Chowder   *Cherrystone*   *Littleneck*

Varieties	Origin	Edible Portion, %
Hard Clams (quahog)		
Chowder (large)	New England	14–20
Cherrystone (medium)	Chesapeake Bay	7–8
	Middle Atlantic	6–8
Littleneck (small)	Pacific Coast	24–28
	Alaska	
Soft shell clams (steamers)	New England	23–33
	Middle Atlantic	27–32

Weight: Varies according to type

Market form: Live in shell, shucked, fresh, frozen, frozen breaded raw or fried, canned, whole or minced

Use: Littleneck and cherrystone—raw on the half shell, stuffed and baked. Quahogs—chowder or clam cakes, breaded strips (Littlenecks are sometimes used in place of soft shell clams as steamers.)

Best buy month: Soft shell—June through September
Hard shell—April, May, June, October, November, December

Purchasing unit: Chowder (large size)—125-150/bushel; shucked 100-250/gallon
Cherrystone—250-300/bushel; shucked 131-188/gallon, 82-131/gallon (medium)
Littleneck—500-600/bushel, shucked, more than 188/gallon

DESIRABLE CHARACTERISTICS

In Shell:
1. Should be alive (neck of soft shell will move).
2. Shell should be closed or close tightly when gently tapped.
3. When two clams are tapped together, they should sound like hard pebbles.

Shucked:
1. Plump flesh with creamy color.
2. Clear liquor.
3. Free from shell particles.

UNDESIRABLE CHARACTERISTICS

In Shell:
1. Open shell that remains open when pressed (indicating that the clam is dead).
2. Meat is dried up, no liquor.
3. Meat may be discolored, soured, and have bad odor.

Shucked:
1. Meat is dried up and discolored.
2. No liquor.
3. Bad odor.

Note: Clams are bivalves. They filter out microorganisms from the water and as a result, if in areas where pollution is heavy, can carry viruses such as hepatitis, which can be transmitted to man if the clam is consumed raw. For this reason, most state health codes require clam tags to be provided to the food service operator by the wholesaler. These tags list the waters from which the clams were dug and identify the persons who harvested the clams. This is the restaurateur's proof that the clams were purchased through an authorized dealer from safe water.

CLEANING AND SHUCKING CLAMS   Clams live in the sand or mud under water. A certain amount of sand or mud still adheres to the surface of clams when they are delivered to the food operation. It should be washed off with clean fresh water, and, if necessary, the clams should be brushed or wiped individually to remove excess grit. The clams should then be placed in an uncovered container. No water or ice should be put over the clams. The clams, still living, will, over a period of a few days, spit out a certain amount of liquid, which will collect at the bottom of the container. Sand and grit present within the clam is discharged by this self-cleaning process and accumulates in the liquid at the bottom of the container.

When the time comes to open the clams for service, the clams should be handled very gently. If they are handled abruptly and bounced around, they will tighten up and become very difficult to open. Many tales have spread throughout the food service industry about quick and easy ways to open clams. Some operators soak clams in club soda and fresh water with corn meal added. None of these methods does anything to decrease the difficulty of opening a clam. If a clam is kept under refrigeration and handled gently, the muscle is actually in a relaxed state, which makes opening the clam very easy. The following step-by-step method should assure success in opening hard shell clams:

Step 1
Locate the indentation on the hinge side of the clam and turn the indentation toward the top.

Step 2
Place the clam (hinge side toward the thumb) in the palm of your hand. Insert the edge of a clam knife between the top and bottom shell.

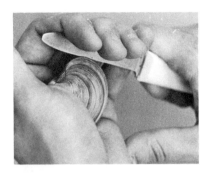

Step 3
Use the fingers of the hand holding the clam to gently press the knife between the two shells.

Step 4
Enter the clam only part way, so that the muscle is not cut in half.

Step 5
Twist the clam knife toward you to pry open the shell slightly.

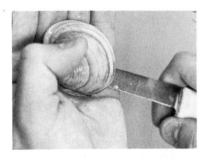

Step 6
Using the tip of the knife blade, run the edge of the knife across the top shell, cutting the clam muscle from that shell and leaving it in the bottom shell.

Step 7
Remove the top shell by prying it up with the knife or thumb.

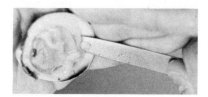

Step 8
Loosen the clam muscle from the bottom shell, being careful not to cut into the muscle. If the clam is to be served on the half shell, leave the muscle in the shell with the clam liquor. Otherwise remove the muscle and prepare it for other uses—for example, chop it for stuffed clams or chowder.

Oysters

Oyster

Oyster on half shell

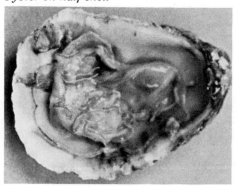

Varieties	Origin	Edible Portion, %
Eastern	New England and Middle Atlantic	8–11
	Chesapeake Bay	6–7
	South Atlantic	4–6
	Gulf of Mexico	5–7
Pacific	Pacific Coast	10–14

Market form:	Live in shell, shucked, fresh or frozen breaded or raw or fried, canned
Use:	Appetizer—raw on the half shell, baked
	Entree—fried, stewed

Purchasing unit:

Live in Shell, Oysters/bushel

Bluepoint	320–400
Medium	200–240
Large	120–160

Shucked

Grade	Oysters/gallon
Counts or extra large	Under 160
Extra selects or large	161–210
Selects or medium	211–300
Standards or small	301–500
Standards or very small	Over 500

Best buy months: Available all year

DESIRABLE CHARACTERISTICS
In Shell: 1. Should be alive.
      2. Shells should be closed or should be closed tightly when gently tapped.
      3. Flesh should be plump, well-shaped, creamy color.
      4. Clear liquor.
Shucked: 1. Plump flesh, well shaped, creamy color.
      2. Clear liquor (no shell particles).
      3. Fresh odor.

UNDESIRABLE CHARACTERISTICS
In Shell: 1. Open shell (oyster is dead).
      2. Meat soft, dried up, or discolored.
      3. Liquor turned opaque to grayish color or no liquor at all.
      4. Sour or offensive bad odor.

Oysters are bivalves, like clams, and should be handled in a similar fashion. They can be scrubbed and cleaned and then put into containers for storage, under refrigerated conditions. Opening oysters is a bit more difficult than clams because the shells have razor sharp edges. Caution should be used in opening oysters, as serious cuts are a real danger.

To begin with, it is a good idea for the food service worker to wear gloves when shucking oysters. An oyster does not have as noticeable a separation as the clam. When opening oysters for the first time, it is advisable to use a pliers or hammer to break the outer tip of the oyster shell. This will expose the separation between the two shells. Once this is done, the procedures listed should be followed:

SHUCKING OYSTERS

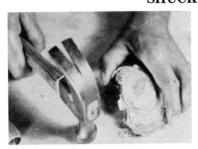

Step 1

Using a pair of pliers or a hammer, break a small section of the tip of the shell off the oyster, exposing the separation between the top and bottom shells.

Step 2

Grasp the oyster in one hand, holding the bowl-shaped shell on the bottom side to reserve the oyster liquor. Insert the tip of the oyster knife into the opening visible and pry the two shells apart.

Step 3

Run the oyster knife across the top or flat shell to separate the muscle from the shell, being careful to contain all of the liquor, with the muscle, in the bottom shell.

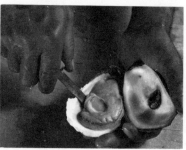

Step 4

Using the knife, cut the small muscle, which attaches the oyster to the bottom shell, so that it is easy for the guest to consume it.

Step 5

The oyster can be opened from the back or hinged side by inserting the tip of the oyster knife between the top and bottom hinge. Pry up with the tip of the knife until the shell lifts slightly. Then proceed as normal with steps 2–4.

CLAM KNIFE

OYSTER KNIFE

Scallop

Varieties	Origin	Size of Eye Muscle	Shucked scallops/lb
Bay (more tender)	New England, Middle and South Atlantic, Gulf of Mexico	½ in. across	40
Sea (less tender)	New England, Middle Atlantic	2 in. across	10–15
Market form:	Shucked, fresh, frozen, frozen breaded raw or cooked		
Use:	Fry, broil, sauté		
Purchasing unit:	*Shucked Meats (count/gallon)*		
	Bay 250–350		
	Sea 100–150		
Best Buy Month:	Bay—October through March		
	Sea—April through October		

DESIRABLE CHARACTERISTICS

Color:	Bay—creamy white to slightly pink color
	Sea—white color
Odor:	Sweet, almost nutty
Texture:	Delicate, bay smoother than sea scallop

UNDESIRABLE CHARACTERISTICS

Color:	Browns as it ages, deathly white if drenched with ice water for too long a time
Odor:	Not sweet
Texture:	Coarse

Lobster

Species:	Lobster Americanus
Origin:	New England, Canada, Nova Scotia
Type:	Chicken 1 lb
	Quarters 1¼ lb
	Selects 1½–2½ lb
	Jumbo 2½–3½ lb
	Cull lobster with one claw or claw missing
Market form:	Live in shell, frozen in shell, canned meat, cooked frozen meat
Use:	Boil, broil, lobster Thermidor, lobster Newburg, bake stuffed
Best buy months:	June through October
Purchasing unit:	Live in shell: 1–4 lb (some lobsters grow to 25 lb)
Proportion edible:	20–35%

DESIRABLE CHARACTERISTICS

Live:
1. Heavy hard shell.
2. Color of shell should be mottled bluish-green.
3. Claws and tail should show movement to indicate lobster is alive.
4. Both claws should be intact.
5. Tail should curl under body when lobster is picked up.

Cooked:
1. Heavy for its size.
2. Color of shell should be bright red.
3. Odor should be desirable.
4. Tail should curl firmly under body.
5. Flesh should be firm and rigid.
6. Tail should spring back into place once straightened.

UNDESIRABLE CHARACTERISTICS

Live: Darkening of the meat caused by bruising or leakage of blood into the tissues.

Cooked:
1. Color is off-red.
2. Tail does not spring back.
3. Flesh not firm, watery with sections digested.

Note: Lobsters must shed shell to grow. When in the canvasback stage (after shedding and reforming a new shell), there is virtually no meat in the shell. Lobster shells at this point may be full of water and may almost explode when cut into. Return shedders to supplier.

BOILING LOBSTER If the lobster is to be boiled, it should be live prior to preparation. Enough water should be brought to a rolling boil to cover the lobster or lobsters. After plunging the lobsters into the water, cover the kettle, bring back to a boil, and cook until done. Cooking time varies with the size of the lobster and the number of lobsters in the pot. A 1–1¼ lb lobster cooks in about 7 minutes.

PREPARING A LOBSTER FOR BAKED STUFFED OR BROILED LOBSTER

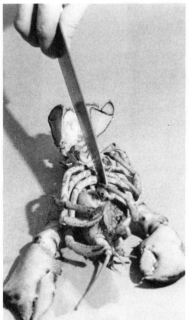

Step 1
Place the live lobster on its back. Using a sharp French knife or lobster splitting knife, cut from the head down through the tail without separating the shell along the back.

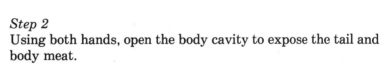

Step 2
Using both hands, open the body cavity to expose the tail and body meat.

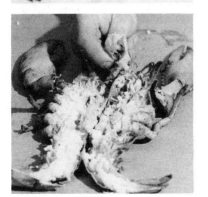

Step 3
With the fingers, remove sac and intestinal tract. (**Note:** the sac is located behind the eyes and is attached to the long digestive tract that runs the length of the body and tail.)

The lobster is now ready to be stuffed. The tamale, a green-colored mass located inside the body cavity, which has a very distinctive flavor, should be mixed with the stuffing. It is also recommended that the small feeler claws or legs be removed and laid across the stuffing to prevent it from overbrowning.

Many restaurants use a lobster rack when broiling a lobster to prevent the tail from curling. If a rack is not available, the claws can be removed from the body and laid over the tail. This should be enough weight to prevent the curling.

A lobster is done when the translucent color of the flesh disappears and becomes pure milk white. The meat in the tail section begins to pull away from the shell slightly and develops a firmness to the touch. Many people overcook lobsters, which causes the flesh to toughen and dry out. A 1¼-lb baked stuffed lobster should be baked for approximately 10 minutes and finished under a broiler for 2 to 4 minutes.

To distinguish a male from a female lobster, turn the lobster over on its back and look at the juncture where the tail meets the body. There are two relatively small legs or flippers at this location. The female's are very soft, delicate, and flexible and quite often cross over, as indicated in the photo. The male's are larger, coarser, and tend to run straight rather than cross.

The only purpose for distinguishing female from male is to select a lobster carrying roe, which is the essence of lobster flavor and is often considered a delicacy. The roe is distributed along the lower body and tail of the female. In the raw state, it is a dark deep green, but when the lobster is cooked it becomes a brilliant orange-red. As for the flesh itself, there is virtually no difference between the male and the female.

FEMALE LOBSTER

MALE LOBSTER

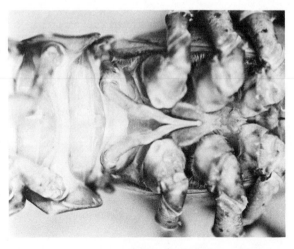

Lobster Tail

Species:	Spiny or Rock Lobster
Origin:	1. Cuba, Florida, Bahamas—brownish-green shell with white spots
	2. South Africa, Australia, New Zealand—dark maroon to brown rough shell.
	3. Southern California to the West Coast of Mexico: yellow-green smooth shell.
Market form:	Frozen in shell, canned meat, cooked meat, fresh
Cooking method:	Broil, bake
Purchasing unit:	Live in shell, 1–4 lb; Headless, raw, frozen, 2 oz to 1 lb.

Shrimp

1. *Live or green shrimp*
2. *Headless shrimp*
3. *Peeled shrimp*
4. *Veined shrimp*

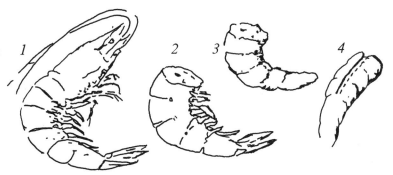

Varieties	*Origin*	*Headless raw, proportion edible*
White, brown, pink	South Atlantic, Gulf of Mexico	47–55%
Alaska pink	Alaska	47–55%
California Gray	Maine, Mexico, French Guinea, Dutch Guinea	47–55%
Market form:	Live or green shrimp—Shell and head on. "Green shrimp" means raw shrimp, not the color of the shrimp.	

Market form:

Green headless—head removed. Shell is intact.

Peeled shrimp—headless shrimp with shell removed.

Peeled and deveined—headless shrimp, peeled and black digestive canal removed.

Peeled, deveined, and individually frozen (PDIF)

Peeled, deveined, quick frozen (PDF).

Individually quick frozen (IQF).

Frozen breaded round raw or fried—deveined and breaded.

Breaded butterfly—deveined, split down the center but not completely through, flattened out, and breaded.

Cooked—peeled, deveined, cooked, and frozen.

Frozen canned

Dehydrated

Broken shrimp

Use:

Appetizer—shrimp cocktail, shrimp salad, barbecued shrimp

Entrees—shrimp scampi, fried shrimp, baked stuffed shrimp, shrimp tempura, shrimp Newburg, creole, curry, paella.

Best buy months. May through November.

Purchasing unit:

Green Headless Count/lb:		Peeled and Deveined Count/lb*	From Green Headless size
Under 10	Average		
10–15	13		
16–20	18	22–25	16–20
21–25	23	29–31	21–25
26–30	28	35–37	26–30

*Approximately 25% is lost during the shelling and deveining process. Once shell and tail are removed, more shrimp must be added to make up the pound weight. So 16–20 green headless size when peeled and deveined yield the size shrimp that are 22–25 count per pound.

Breaded

(Count/lb includes breaded weight, i.e., 16–20 = 18/lb)

Regularly breaded-frozen raw breaded shrimp: minimum 50% shrimp.

Lightly breaded-frozen raw breaded shrimp: minimum 65% shrimp.

Any shrimp that contains over 50% bread crumbs must be labeled imitation shrimp.

Cooked Shrimp

Sold by the pound.

Cooked Meat in Cans

Size container: 6-, 8-, or 12-oz- or 1-5-lb tin.

DESIRABLE CHARACTERISTICS

- Fresh shrimp should slip crisply over one another.
- Mild sea odor.
- Firm texture.
- Semitransparent grayish-green color.
- Flesh should adhere to shell.
- If frozen, a good, even glaze cover should protect shrimp from freezer burn.

UNDESIRABLE CHARACTERISTICS

- Ammonia odor or high iodine content.
- "Black spot" present, causing shrimp to deteriorate, reducing quality, and causing a reddish color (not to be confused with the color of good-quality grooved shrimp and Royal Reds).
- Sliminess due to improper storage, producing a strong off-taste and off-odor.
- In frozen shrimp, dehydration due to poor packaging, causing a moisture loss and damaging the texture of the product.

PREPARING SHRIMP SCAMPI

1. Using large green headless shrimp, peel off the shell down to last section but leave on the tail.
2. With knife, cut along the curved back and remove the digestive tract.
3. Butterfly the shrimp and place them on broiler pan or in sauté pan.
4. Add garlic butter and sauté or broil just until translucent color disappears and shrimp become pure white.

PREPARING BAKED STUFFED SHRIMP

1. Use large green shrimps, U–15 count. Remove legs and cut down belly side to digestive tract.
2. Remove intestinal vein; then loosen meat from shell.
3. Fill shrimp with stuffing made from bread crumbs, crab meat, and seasoning.
4. Bake approximately 10 minutes at 350°F. Finish under broiler for 2 minutes.

PREPARING SHRIMP COCKTAIL

Green headless PDQ or IQF shrimp may be used. Most suitable sizes are U–15, 16–20, or 21–25 count per pound.

1. Bring water to boil.
2. Submerge shrimp in boiling water. If shrimp are thawed or fresh, they should be done in 3 to 5 minutes.
3. If frozen, start timing from the time the water returns to boil and shrimp separate from block.

Note: Shrimp are often overcooked, causing a rubbery texture and dry flavor. Shrimp are done when the color turns milky white. It is important to ice shrimp immediately when they have reached this point or they will overcook in the hot water.

Crab

BLUE CRAB

DUNGENESS CRAB

KING CRAB

ROCK CRAB

Crab photographs courtesy of the National Oceanic and Atmospheric Administration (NOAA). Used with permission.

DESIRABLE CHARACTERISTICS
Live in Shell:
1. Should show movement of the legs.
2. Should have good sweet odor.
Cooked in Shell
1. Should be bright in color.
2. Should be free from bad odors.
Crab Meat
1. Should have sweet odor.
2. Should contain yellow fat particles in the white meat.
3. Should not have chemical preservatives or grit.

UNDESIRABLE CHARACTERISTICS
1. Sour ammonia odor.
2. Slimy or sticky flesh.
3. Mushy flesh

Variety	Origin	Weight, lb	Market Form	Use
Blue crab Hard shell	Atlantic, Gulf Coast	1/4–1	Live, cooked in shell, fresh cooked, meat, frozen cooked meat, canned	Cooked whole, lump, and flake meat removed from body cavity and claws is the most desirable type of crab meat available. Used for crab cocktail, crab Newburg, crab meat au gratin, crab meat salad, and other dishes.
Soft shell		1/4–1	Live, frozen uncooked	Generally prepared whole, either sauteed or deep fried.
Dungeness	Pacific Coast	1¾–3½	Live, cooked in shell, fresh cooked meat, frozen cooked meat, canned	Used the same way as the blue hard shell crab.
King	Alaska	6–20	Frozen cooked in shell, frozen cooked meat	Body meat is used for crab salad, crab meat *au gratin*, crab Newburg. The claws and legs are generally cut into sections, sold precooked, frozen, and used for broiled crab legs. In many cases, it is substituted in place of lobster tails. The meat is firm and chunky.
Snow	Alaska	3–10	Same as King crab	Similar to King crab but less firm. Size of the legs and claws is smaller, so they do not lend themselves to being used as a broiled item.
Rock	New England, California	1/3–1/2	Live, fresh cooked meat, canned meat	Same use as the blue hard shell crab.
Stone	Florida	2–6 oz (Only claws and claw meat are sold.)	Cooked in shell, frozen in shell, cooked meat	Claws are sold precooked. Stone crab claws can be used as an appetizer, cold, or an entree steamed and served with drawn butter. They are also used in many cases as a substitute for higher priced lobster.

REMOVING MEAT FROM BLUE CRABS

Step 1
Remove claws from whole crab. Crack claws and remove meat.

Step 2
Insert knife or finger under back side of top shell. Remove top shell by prying down with knife or lifting up with finger.

Step 3
Remove internal organs and digestive tract from body cavity. Rinse out with fresh water.

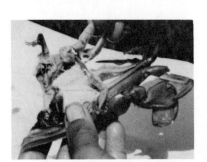

Step 4
Break shell in half and separate fibrous membrane from body.

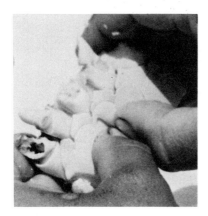

Step 5
Break each section in half again and remove body meat from membrane holding it to shell.

Mussels

Origin:	New England, Middle Atlantic, Canada. Also commercially cultivated in Maine, Atlantic Coast of France, the Netherlands, and Spain.
Market form:	Live in shell, breaded quick frozen, shucked and frozen, canned, smoked, and pickled.
Uses:	Steamed, *moules marinières*, in bouillabaisse
Purchasing unit:	Live in shell, 55 lb/bushel

DESIRABLE CHARACTERISTICS
1. Should be alive in shell.
2. Shells should be closed or close tightly when gently tapped.

UNDESIRABLE CHARACTERISTICS
1. Open shells that remain open when pressed.
2. Meat discolored or foul odor.
3. Filled with mud. (Slide two halves of the shell across each other; if they move, shell is probably filled with mud).

CLEANING Mussels have a "beard" that has to be scraped off. This can be accomplished by using a stiff brush, by clipping the beard with scissors, or by simply pulling it off with the fingers. The mussels should be washed under cold running water.

Squid

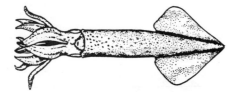

Other Name:	Calamari
Origin:	Atlantic Coast, Pacific Coast, Gulf of Mexico, imported
Market forms:	Fresh, frozen
Cooking method:	Fry, bake in sauce
Average size:	10–12 in. long

Other Seafood Items And Their Uses

SNAILS Periwinkles are used as appetizers. The French word *escargot* is used for snails served in the shell, cooked in garlic and herb butter.

ABALONE Steaks, slivers in soup

FROG LEGS *Provençale,* deep fried

10.6 NEW PRODUCT—MINCED FISH

As mentioned earlier, there are over 50,000 varieties of fish in the sea but only 240 species are currently marketed. Many species are not marketed because it is difficult to fillet or debone the fish yet there is customer resistance to buying the fish whole, because of its appearance. Many species that are perfectly edible are routinely killed and dumped back into harbors, bays, lakes, and rivers when caught by commercial fishermen, who believe there is no market for the less known species.

Mechanical devices have been designed in the past to fillet certain species of fish. These devices generally leave anywhere from 20 to 40 percent of the meat on the backbone. Although perfectly edible, this meat generally goes into the trash barrel or possibly into fertilizer. Food scientists throughout the United States, trying to recover the usable portion of this meat, have developed a number of methods to remove it from the backbone. Probably the most successful method, which can be used on any species that has been gutted, is that of the mechanical deboning machine called the Beehive Deboner (see Figure 10.1). This device grinds up all of the flesh with the skin and bone and by the use of a screw extractor, extrudes the meat through very tiny orifices, leaving the bone and skin behind. The end product is 100 percent edible and looks similar to a finely ground beef product with the exception of the color.

This device can be used on products that are currently considered unmarketable—for example, many species of lake fish such as mullet and sucker and ocean fish such as sea robins and shark. In 1980 the cost per

Figure 10.1   The beehive deboner.

pound for completely usable minced fish was approximately 60¢. The following are some of the menu items that have been prepared from this minced fish product: fish chowder, fish "crispies" or fritters, fish hot dogs, lasagna, tacos (see Figures 10.2–10.5).

The minced fish would be excellent to use in chowders and in items such as baked stuffed clams, bouillabaisse, Newburg casseroles, and creoles, with substantial savings compared to using whole fish or fish fillets.

Figure 10.2   Packaged minced fish.

Figure 10.3   Chowder made from minced fish.

Figure 10.4   Hot dogs made from minced fish.

Figure 10.5   Fish crispies—fritters made from minced fish.

Technological advances should make fish and other seafood products more and more attractive to the food service operator. With a good working knowledge of new products, the food operator can offer his customers quality, variety, and economy on his food service menu.

10.7 SUMMARY

Although fish and shellfish may never reach the consumption levels achieved by red meats or poultry in the United States, they are very important items on many food service menus today. They vary from the economical (the fast food fish sandwich) to the extravagant (baked stuffed lobster). At the same time fish products provide high-quality protein and relatively low levels of fat.

The 200-mile limit combined with advances in the technology of fish farming may provide the food service industry with additional supplies and species of fish and shellfish to market in the future. The information contained in this chapter should provide a basis for understanding and utilizing fish and shellfish on the food service menu.

CASE STUDY

1. Ezra's Fish House on the Wharf in Boston features the finest in fresh local seafood. The following list is just a sample of the items offered on the menu:
   - New England Clam Chowder
   - Baked Clams Casino
   - Oysters on the Half Shell
   - Jumbo Shrimp Cocktail
   - Fillet of Sole
   - Cod and Haddock
   - Shrimp Scampi
   - Fried Shrimp
   - Baked Stuffed Lobster
   - Soft Shell Crab

   Discuss the products purchased to prepare the menu items, their market form, size, condition, and so on. What quality attributes should each item possess upon delivery and what warning signs might warrant refusal of the products?

2. Sorry Charlie's Fishery delivers your order, and your receiving clerk makes the following list:

   Fresh rainbow trout (dressed)—loose scales, sunken eyes, soft flesh
   Frozen Dover sole (dressed)—partially thawed, partially glazed
   Frozen salmon steaks—firm flesh, completely frozen
   Clams (live in shell)—alive, closed shells, plump meat
   Oysters (shucked)—no liquor, discolored
   Bay scallops (shucked)—slightly brown, coarse texture

Lobster Americanus (live in shell)—mottled bluish-green shell, tail curls under body

Fresh shrimp (green headless)—reddish color, ammonia odor

Mussels (live in shell)—shells remain closed when gently tapped

Which items should be accepted? Which should be refused and why? How should they be stored and for how long?

11 SOY PRODUCTS

From 1970 to 1980 the price per pound of ground beef doubled. As the cost of even the cheapest cuts of beef rose, individual and commercial consumers looked for substitutes. As fish and poultry have become too expensive to be practical substitutes, vegetable protein, primarily soy, has come into wide use.

Soy, a high-protein substance derived from the seed of soybeans, is becoming a more acceptable substitute in products that were previously prepared with 100 percent meat. New commercial product lines incorporating soy are being developed daily. Already on the market are chicken, fish, and beef hotdogs extended with soy; turkey and chicken loaves and rolls extended with soy; fish fillets with minced fish (using underutilized species such as mullet) and soy; bacon and bacon bit substitutes composed of soy; pizza toppings; and a nondairy, nonmeat Soy Stroganoff. Soy is also being used in recipes for beef, veal, pork, and chicken patties, in chili con carne, taco fillings, sausages, meat loaves, and a variety of other dishes. Its potential use is much more extensive.

11.1 MARKET FORMS OF SOY

Soy is available in a variety of forms from a number of different manufacturers (see Table 11.1). The most common forms in the marketplace are, ranging from least to most expensive,
1. Soy flour and grits. Soy flakes ground to a fine powder.
2. Soy concentrates. Soy flakes treated to remove some of the nonprotein components (that is, fiber and carbohydrates).
3. Soy isolates. Soy flakes treated to isolate the protein.

The protein is lowest in soy flour and highest in soy isolates.

Fully fatted soy flour is approximately 40% protein on a moisture-free basis (the moisture is removed from the bean), compared to defatted soy flour, which is about 50% protein. As more of the nonprotein components are removed, the protein level increases. The protein content of soy concentrates ranges from 60 to 70%, and the content in soy isolates goes as high as 90%. Textures can also be added to the different forms. Although textured vegetable proteins cost more than nontextured vegetable proteins, the result is an end product that more closely resembles the all-meat product.

11.2 ADVANTAGES OF SOY

Soy has three major advantages: it is versatile, nutritious, and economical. Advances in technology are providing increased production, more efficient processing, and better palatibility for soy products.

Versatility of Soy

Soy products have the ability to blend very well with other ingredients. Soy is available uncolored to blend with fish, veal, pork, and poultry items or caramel colored to simulate the color of cooked meat. When appropriate quantities are added to beef, fish, and poultry preparations, the appearance and taste of the meat or fish remains basically unchanged.

The three major forms of soy are used in a number of ways (see Table 11.1). Soy flour is most often used in manufactured baby foods and baked products. It is also used in formulations for frankfurters, sausages, and bologna.

The food service industry uses the second form, soy concentrate, more often than soy flour. Textured concentrates are successfully used as extenders of chopped (comminuted) meats, used in chili, meatloaf, and meat patties.

Textured soy isolate, the third form, is generally used as a meat analogue to resemble different meat, fish, and poultry products in appearance, flavor, color, and texture. Common examples of soy analogues are bacon bits and slices, "ham" or "chicken" cubes, and simulated luncheon meats.

Nutritional Advantages

Soy is high in protein, low in fats, and contains virtually no cholesterol. Although it is low in methionine, one of the eight essential amino acids, nutrients can be added to the soy product to enhance its nutritional properties. When used in combination with meat or other vegetables high in methionine, the proteins complement each other and the nutritional value of soy increases to that of a complete protein.

In its dry form, the percentage of protein in soy (ranging from 50 to 90%) is very high. When soy is hydrated, the protein percentage is, of course, diluted. Textured vegetable protein product that is 52% protein in dry form, when hydrated with two parts water to one part soy, yields a ready-to-use product containing 17% protein.[1] But the product still compares favorably with an all-beef product, which has an average protein content of 15–20%.

Economic Benefits

Pound for pound, soy is less expensive than meat. As illustrated by Table 11.2, soy protein is a fraction of the price of red meat protein. The 1979 prices listed are based on the price for a 50-lb bag of the soy product. Soy purchased in larger quantities, say 25,000 lb, would be priced 7–10¢/lb

TABLE 11.1   FUNCTIONAL PROPERTIES OF SOYBEAN PROTEIN IN FOOD SYSTEMS

Functional Property	Protein Form Used*	Food System
Emulsification		
Formation	F,C,I	Frankfurters, bologna, sausages, breads, cakes, soups, whipped toppings, frozen desserts
Stabilization	F,C,I	Frankfurters, bologna, sausages, soups
Fat absorption		
Promotion	F,C,I	Frankfurters, bologna, sausages, meat patties
Prevention	F,I	Doughnuts, pancakes
Water absorption		
Uptake	F,C	Breads, cakes, macaroni, confections
Retention	F,C	Breads, cakes
Texture		
Viscosity	F,C,I	Soups, gravies, chili
Gelation	I	Simulated ground meats
Chip and chunk formation	F	Simulated meats
Shred formation	F,I	Simulated meats
Fiber formation	I	Simulated meats
Dough formation	F,C,I	Baked goods
Film formation	I	Frankfurters, bologna
Adhesion	C,I	Sausages, lunch meats, meat patties, meat loaves and rolls, boned hams, dehydrated meats
Cohesion	F,I	Baked goods, macaroni, simulated goods
Elasticity	I	Baked goods, simulated meats
Color control		
Bleaching	F	Breads
Browning	F	Breads, pancakes, waffles
Aeration	I	Whipped toppings, "chiffon" mixes, confections

Source: W.J. Wolf, Soybean Protein: Their Functional, Chemical, and Physical Properties, *Journal of Agricultural Food Chemistry* 18 (1970): 971.
*F,C,I designate flours, concentrates, and isolates, respectively.

TABLE 11.2   COST COMPARISON BETWEEN SOYBEAN PRODUCTS AND GROUND BEEF*

Product	Cost/lb, $	Appropriate Protein, %	Cost/lb. of Protein, $
Defatted soy flour	0.25	50	0.50
Soy protein concentrate	0.45	70	0.64
Soy textured concentrate	0.60	50	1.20
Soy protein isolate	1.00	90	1.11
Ground beef	1.50	18–20	7.50

*1979 prices based on 50-lb bags. As the price of ground beef continues to escalate, the economic feasibility of extending ground beef based menu items with soy becomes more apparent.

lower. (Note that prices are never constant and are quoted here only for illustrative purposes.

11.3 STEPS FOR SUCCESSFUL USE OF SOY IN FOOD SERVICE

Selecting the Soy Product

Begin with a high-quality soy product. Since the soy product is so much cheaper than the ground beef, it does not pay to skimp on the quality of the soy to save a few cents per pound. For extending ground beef, purchase a soy concentrate that has had a large portion of the carbohydrates removed (manufacturers' specification sheets contain information on carbohydrate content). This product has no bitter, beany flavor and does not cause digestive tract disturbances in customers as beans often do. When purchasing a meat analogue such as bacon bits select a brand name and make taste comparisons.

Deciding How Much to Use

Do not extend meat beyond reasonable limits. In general, the lower the percentage of soy, the better the overall palatability. An unseasoned item, like hamburger, should not be extended more than 10 to 20 percent with soy. Beyond this limit, very noticeable flavor and texture changes occur that make the end product less desirable. Some products that are normally highly seasoned, like chili and tacos, can be extended using higher levels of soy than recommended for hamburger. As much as 20, 30, and even 40 percent can be used with very good results with the proper procedures. Recipes with recommended percentages and procedures are available from the manufacturers. To incorporate more soy, hydrate the soy in beef stock rather than plain water. (Hydration time ranges from 10 minutes to 1½ hours.) This will give the finished product a flavor more closely resembling that of an all-beef product. Keep in mind your market and menu price, extending only to the limits appropriate for both.

Food service managers must convince their employees of the nutritional advantages and economy of soy. Like any new product it will probably be looked at with skepticism. Employees must have a positive attitude toward any product they are required to sell. If they are not informed and are not given the opportunity to sample the soy products, they will most likely discourage customers from ordering menu items containing soy.

The percentage of soy to be used depends on the product, the type of operation, the menu price, and other factors. Each individual food service operator must decide how much to use. Hamburgers, meat loaf, chili, lasagna, tacos, Sloppy Joes, and meatballs are but a few of the items that successfully include soy as an extender.

In the following examples one can see the potential profitability of using soy in popular dishes like chili, meat loaf, and hamburgers. Because market prices as well as percentages of soy used will vary from operation to operation, these examples are for illustrative purposes only.

The following analyses compare costs of a batch of chili, a meat loaf, and hamburger made with 100% ground beef and a batch extended with 40%, 30%, and 20%, respectively, of hydrated soy.

Cost of ground beef	= $1.50/lb
Cost of soy concentrate	= $0.45/lb
Cost of soy hydrated (2 lb water to 1 lb soy)	= $0.15/lb

CHILI

100% ground beef
100 lb ground beef × $1.50/lb = $150.00

60% ground beef, 40% hydrated soy
60 lb beef × $1.50/lb = $90.00
40 lb hydrated soy × $0.15/lb = $6.00
$90.00 + $6.00 = $96.00
Results:
$150.00 (100% ground beef) – $96.00 (60% ground beef + 40% hydrated soy) = $54.00 savings per batch of chili.

MEAT LOAF

100% ground beef
100 lb ground beef × $1.50/lb = $150.00

70% ground beef, 30% hydrated soy
70 lb beef × $1.50/lb = $105.00
30 lb hydrated soy × $0.15/lb = $4.50
$105.00 + $4.50 = $109.50
Results
$150.00 (100% ground beef) – $109.50 (70% ground beef, 30% hydrated soy) = $40.50 savings per batch of meat loaf.

HAMBURGER

100% ground beef
100 lb ground beef × $1.50/lb = $150.00

80% ground beef, 20% hydrated soy
80 lb beef × $1.50/lb = $120.00
20 lb hydrated soy × $0.15/lb = $3.00
$120.00 + $3.00 = $123.00
Results
$150.00 (100% ground beef) – $123.00 (80% ground beef, 20% hydrated soy) = $27.00 savings per 100 lb of hamburger

Lower Cost Meat Formulation

It is important to note that when extending ground beef with soy, fatter ground beef should be used than would make good all-meat items. This is because the soy protein is virtually fat free. If the desired product is to have a lean-to-fat ratio of 80 to 20, the ground beef must contain proportionately more fat; recommended ratios of lean to fat are 70:30 or 60:40. Following this practice also results in savings since fattier ground beef is less expensive.

In addition to dollar savings, tests show that cooked beef extended with hydrated texture vegetable protein retains more water than all-beef patties. Consequently, the extended patties are not only moister than all-beef patties, but their ability to hold water reduces the shrink loss that usually accurs during the cooking of all-beef products.

11.4 STORAGE OF SOY

Shelf Life

In its dry form, soy has an excellent shelf life: it will last about one year. Once hydrated, it must be handled the same as fresh meat. It should be cooked, refrigerated, or frozen within a specified period of time.

Under refrigeration the shelf life of hydrated soy and of ground meat extended with soy varies depending on the initial quality and "plate count" (the number of bacteria) of the meat as well as on the temperature and the sanitation conditions existing in the storage facilities. Refrigerated shelf life of soy-extended meats is somewhat shorter than that of all-meat products. A study done by Keeton and Melton[2] using separate batches of ground beef (25% fat) extended with textured soy protein at various levels (0 to 30% soy), demonstrates that spoilage bacteria multiply more rapidly in soy-extended ground beef than in the nonextended control. In this study, samples were held at 5°C (41°F) for four days. At the end of this period, spoilage was evident in the soy-extended samples. The study cites the primary reason for this increased bacterial growth to be the higher pH (6.5) of the soy-extended ground beef. This environment provides a much more favorable condition for bacterial growth than the normal pH of 5.5. With this in mind, refrigerated storage of soy-extended ground beef should be controlled carefully. If temperatures can be kept at 0°C (32°F), bacterial growth can be slowed and the shelf life can be extended. At 32°F it takes an average microorganism 38 hours to double; at 50°F it takes only five hours. If storage temperatures average 38–40°F, then the soy-extended ground beef should be used within three days.

Frozen Storage

Soy-extended ground beef patties can be stored at least as long as ground beef alone. Kotula[3] evaluated frozen beef patties extended with soy during a 12-month period. The patties, containing 20 to 30% soy concentrate and 20% fat, were stored at 0°F and evaluated trimonthly for one year. The results show that the mean scores for flavor, aroma, and overall acceptability did not significantly change during the twelve-month period. Some differences in tenderness, appearance, and juiciness were observed during the year. The study also concludes that soy protein may have some ability to inhibit the development of rancidity.

In the frozen state, soy-extended beef can be handled the same as beef alone. The recommended good-quality storage life for soy-extended ground beef with proper packaging (see Chapter 16, Storage, for details) is four months at 0°F, six months at –10°F, and eight months at –20°F.

11.5   USERS OF VEGETABLE (SOY) PROTEIN

The school systems are among the largest users of soy products in the United States (see Table 11.3). In 1971, soy was approved by the USDA for the Type A school lunch program because of its nutritional properties and cost savings compared to all meat products. In accordance with government regulations, the proportion of hydrated vegetable protein product to uncooked meat cannot exceed 30 parts soy per 70 parts meat on the basis of weight. Since 1971 schools have been quick to incorporate soy in many of their luncheon items, including hamburgers, meat loaf, tacos, chili, barbecue beef, spaghetti sauce, pot pies, casseroles, and ham and chicken salads.

Colleges and universities, nursing homes, and hospitals rank second, third, and fourth, respectively, as the next largest users. All use soy extenders and meat analogues to reduce food costs while at the same time providing palatable meals to their consumers.

Commercial institutions—that is, fast food restaurants, full-service restaurants, and hotels—use soy in limited quantities. Full-service restaurants featuring all natural foods or vegetarian entrees are currently marketing soy as a health or diet menu item for nutrition or weight conscious customers. Some fast food operations are using soy primarily for pizza toppings or highly seasoned foods, as in Mexican dishes.

In May 1979 the U.S. government approved the use of soy for the military. For every 100 lb of product, the military can use 80 lb of beef with 20 lb of hydrated (2.6:1 hydration ratio) granular soy concentrate (70% protein level).

Soy Regulations

The U.S. government regulates the percentage of soy allowed in the school lunch program and in military feeding. The USDA also regulates

TABLE 11.3   USE OF VEGETABLE (SOY) PROTEIN

	Total	Commercial Fast Food	Full Service	Hotel/ Motel
Yes, %	25.7	17.5	10.3	4.1
No, %	74.3	82.6	89.7	95.9

	Hospital	Institutional Nursing Home	School	College/ University	Empl. Feed	Transp.
Yes, %	22.4	35.8	64.4	43.9	13.9	12.5
No, %	77.6	64.2	35.6	56.1	86.1	87.5

Source: User Acceptance Spurs Demand for Vegetable Protein Analogs, *Institutional Volume Feeding Convenience Casebook,*vol. 79:7.

the percentage of soy allowed in processed products if the product is identified by a product name for which a standard of composition has been established. Chili, Salisbury steaks, meatballs, hot dogs, pork sausages, and bologna are examples of such standardized processed products. The level of soy permitted ranges generally from 2% to 12% (see Table 11.4).

If more than the maximum amount of allowable soy is used, the processor must give the product a nonspecific name. Patties and luncheon loaves, such as pepper loaf and spiced luncheon loaf, are such nonspecific products. Proper labeling is required to inform the consumer of the nature of the product.

A truth-in-menu law is pending whereby restaurants may also be forced to inform the consumer if menu items contain soy. They will have to describe soy items ("Beef patty: A nutritious blend of ground beef and soy, seasoned with spices and broiled to perfection").

TABLE 11.4   LEVEL OF SOY PERMITTED IN MANUFACTURED MEAT PRODUCTS*

Manufactured Product	Soy Product	Permitted level, %
Sausage	Flour or grits	3.5
	Concentrate	3.5
	Isolate	2.0
Chili	Flour or grits	8.0
	Concentrate	8.0
	Isolate	8.0
Meatballs for spaghetti	Flour or grits	12.0
Salisbury steak	Concentrate	12.0
	Isolate	12.0

*At these levels the manufacturer does not have to change the name of the product.

TABLE 11.1 MANUFACTURERS AND FORMS OF SOY PRODUCTS

Manufacturers	Products*
Archer Daniels Midland Company P.O. Box 1470 Decatur, Illinois 62525	Textured vegetable protein (TVP), defatted soy flour, full fat soy flour, defatted grits, soy protein concentrates, soy fiber
Cargill, Inc. P.O. Box 9300 Minneapolis, Minnesota 55440	Soy flours, soy grits, (soy) textured vegetable protein
Central Soya Company, Inc. 1300 Fort Wayne National Bank Bldg. Fort Wayne, Indiana 46802	Flours, texturized flours, structured concentrate, concentrate, isolate
Dawson Food Ingredients Inc. 1600 Oregon Street Muscatine, Iowa 52761	Grits, flours, texturized flour, isolates, spun fibrils, spun analogue
Far-Mar-Co., Inc. Food Operations Headquarters 960 North Halstead Hutchinson, Kansas 67501	Textured vegetable protein, ham flavored chiplets, bacon flavored chips
General Foods Corporation 250 North Street White Plains, New York 10625	Cereals, meat analogues
Griffith Laboratories, Inc. 12200 S. Central Alsip, Illinois 60658	Soy protein concentrate, structured soy flour (Offers complete food system to restaurant and industrial caterer.)
General Mills, Inc. Protein Division P.O. Box 1113 Minneapolis, Minnesota 55440	Artificially bacon flavored, textured protein bits and chips
Kraft, Inc. Industrial Food Division P.O. Box 398 Memphis, Tennessee 38101	Isolated soy protein
Launaff Grain Company Box 571 Danville, Illinois 61832	Soy grits, soy flakes, soy flour, textured vegetable protein
Miles Laboratories Inc. Morningstar Farms 7123 W. 65th Street Chicago, Illinois 60638	Meat and fish analogues, breakfast strips, breakfast patties, breakfast links, luncheon slices, patties, meatless hotdogs
Nabisco Advantage East Hanover, New Jersey	Textured vegetable protein extenders for meat, fish, and poultry
Ralston Purina Company Checkerboard Square St. Louis, Missouri 63188	Isolated soy protein, structured protein fiber, textured soy flour
A.E. Staley Manufacturing Company 2200 Eldorado Street Decatur, Illinois 62525	Soy flour, soy grits, textured vegetable protein, soy protein concentrate, textured soy protein concentrate

*Specification sheets containing pertinent information on nutrition, application, packaging, storage, hydration, and labeling of soy are available from the different manufacturers. As with any purchase, product comparison is advisable.

11.6 SUMMARY

The future of soy product acceptance is promising. According to a Gallup poll sponsored by the Food Protein Council in 1977 one-third of the 1543 adults interviewed believed soy to be an important source of protein for the future and believed that the addition of soy increases the overall quality of the product. More than one-half believed soy adds greater nutritional qualities to the product, and more than two-thirds believed that soy blends result in economic savings.[4] With the onset of the school lunch program, younger age groups have accepted the use of soy products. It is only a matter of time before other age groups become aware of its versatility, economy, and nutritional benefits.

As shown by the data in this chapter, soy offers substantial savings in food service operations. Creative use and proper marketing is the key to the success of soy in commercial food operations. When appropriate levels of soy are used to extend meat, the results are excellent. With proper training, kitchen and service staff can overcome any negative feelings they may have toward soy products. Managers of food facilities interested in incorporating soy can conduct taste tests with various menu items prepared with and without soy. Once the staff members taste the results for themselves they will become enthusiastic about the merits of soy products and confident about recommending to the patrons menu items prepared with soy.

CASE STUDY

Llenroc University, like many other universities, offers a special meal plan to the students attending the university. Currently the dining halls are open 235 days a year and serve 16,000 meals per day. Last year the dining halls used 60,000 lb of ground beef and expect to use the same amount this year. Seventy percent of the ground beef was used to prepare hamburgers. The remaining 30% was used to prepare ground beef based recipes, such as lasagna and chili. Last year the price of ground beef was $1.15/lb, and this year the price increased to $1.30/lb. In view of the recent increase in the price of beef and the expected increase in the price for the forthcoming year, the food and beverage manager is considering using a soy concentrate, priced at 55¢/lb, to extend all items prepared with ground beef. Assuming the price of ground beef is expected to reach $1.40/lb next year and consumption of ground beef remains the same, what are the potential yearly savings the manager can expect if the hamburgers are extended with 10% hydrated soy? 20% hydrated soy? 15% hydrated soy? What are the potential yearly savings the manager can expect if the ground beef based recipes are extended with 20% hydrated soy? 30% hydrated soy? 40% hydrated soy? Based on your calculations, what recommendations would you make? What other factors should the food service manager consider before making any decisions using soy as an extender?

REFERENCES

1. Kenneth M. Wolford, "Beef/Soy: Consumer Acceptance," *Journal of the American Oil Chemists Society.* (January 1974), vol. 5:133A.
2. J.T. Keeton, C.C. Melton, "Factors Associated With Microbial Growth in Ground Beef Extended With Varying Levels of Textured Soy Protein," *Journal of Food Science.* (July 1978), 43:1125–9.
3. A.W. Kotula, et al., "Evaluation of Beef Patties Containing Soy Protein During 12 Month Frozen Storage," *Journal of Food Science.* (September 1976), vol. 41:1142–7.

12 | VARIETY MEATS

Variety meats are the liver, brains, kidneys, sweetbreads, heart, tongue, and tripe of beef, veal, pork, and lamb. Rich in protein, vitamins, and minerals (see Table 12.1, a nutritional chart), most are cheaper than most other meat cuts. Sweetbreads, kidneys, and brains command higher prices because they are considered delicacies. Variety meats are highly perishable and should be used soon after purchase. They can be stored in the refrigerator (at 34–36°F for no more than five days or frozen at 0° for no more than three to four months).

Variety meats are often featured on the menus of restaurants of distinction. The quality of their preparation and presentation can contribute to a restaurant's reputation for excellence.

Variety meats are consumed as much for their unique textural characteristics—very different from muscle meats—as for their distinctive flavor. They are enhanced by seasonings and sauces and combine well with other food substances.

12.1 LIVER

Beef: Top
Veal: Center left
Lamb: Center right
Pork: Bottom

Figure 12.1
Livers: 1. Beef 2. Lamb
3. Pork 4. Veal

Liver is rich in protein and iron. It is the variety meat most commonly served in restaurants and schools, hospitals, and other institutions. Beef, veal, and lamb liver are identified by two lobes of unequal size; pork has three lobes of approximately the same size. Calves' liver, the lightest in color, is the most expensive of the four and is prized for its tenderness and mild flavor. Although slightly drier, lamb liver is also mild in flavor and is prepared much the same as calves' liver. Beef liver is dark red in color and is the least tender, followed by pork. Both beef and pork liver have a very strong flavor, which is often masked with onions or bacon.

TABLE 12.1 NUTRITIONAL COMPOSITION OF VARIETY MEATS

	Protein Content of 100-gm Cooked, gm*	Mineral Content of 100-gm Fresh Variety Meats†				
		Calcium, mg	Phosphorous, mg	Iron, mg	Sodium, mg	Potassium mg
Beef						
Brain	11.5	10	312	2.4	125	219
Heart	28.9	5	195	4.0	86	193
Kidney	24.7	11	219	7.4	176	225
Liver	22.9	8	352	6.5	136	281
Tongue	22.2	8	182	2.1	73	197
Tripe		127	86	1.6	72	9
Veal						
Brain	10.5	10	312	2.4	125	219
Heart	26.3	3	160	3.0	94	208
Kidney	26.3	...	...	4.0	...	...
Liver	21.5	8	333	8.8	73	281
Thymus (sweetbreads)	18.4	...	...	2.0	...	...
Tongue	26.2	...	...	3.1	...	...
Pork						
Brain	12.2	10	312	2.4	125	219
Heart	23.6	3	131	3.3	54	106
Kidney	25.4	11	218	6.7	115	178
Liver	21.6	10	356	19.2	73	261
Tongue	24.1	29	186	1.4	...	...
Lamb						
Brain	12.7	10	312	2.4	125	219
Heart	21.7	11	249	...	...	...
Kidney	23.1	13	218	7.6	200	230
Liver	23.7	10	349	10.9	52	202
Tongue	21.5	...	147	3.1	...	...

*Source: Abridged from Lilia Kizlaitis, Carol Diebel, and A.J. Siedler, Nutrient Content of Variety Meats II. The Effects of Cooking on the Vitamin A, Ascorbic Acid, Iron and Proximate Composition. *Food Technology* 18 (1964):103.

†Source: Bernice K. Watt and Annabel L. Merrill, *Composition of Foods—Raw, Processed, Prepared*, U.S. Dept. of Agriculture Handbook No. 8 (Washington, D.C.: U.S. Government Printing Office, 1963).

TABLE 12.2 LIVER FACTS

	Beef	Pork	Veal (Calf)	Lamb
Average weight	10 lb	3 lb	2½ lb	1 lb
Preparation time				
Broiled or sautéed (sliced)	8–10 min	8–10 min	8–10 min	8–10 min
Braised (whole)	2½ hr	1½–2 hr	...	...
(sliced)	20–25 min	20–25 min	...	...
Market form	Fresh, frozen, or canned as liver paste			

12.2 KIDNEYS

Figure 12.2   Kidneys: (left to right) veal, beef, lamb, pork.

Kidneys are located in the pelvic cavity where they are enclosed in a capsule of fat. They are the only organ meat left attached to the carcass after slaughtering. Quite often, veal and lamb kidneys are left attached to the loin during fabrication and sold as veal kidney chops and English lamb chops, respectively. Kidneys may also be purchased separately and are identified as follows: Both beef kidneys and veal kidneys (the lighest in color) consist of numerous lobes, whereas pork and lamb kidneys are bean-shaped. Of the four, the veal kidney is the most tender and delicately flavored, followed by lamb, pork, and finally beef, which is often tough and strong in flavor. Although well liked and frequently prepared as kidney pie or in cream sauces by the Europeans, kidneys are often overlooked by the American consumer. With proper preparation and marketing, this variety meat could appeal to the American palate.

TABLE 12.3   KIDNEY FACTS

	Beef	Pork	Veal	Lamb
Average weight	1¼ lb	¼ lb	¾ lb	⅛ lb
Preparation time				
Broiled or sautéed	...	10–12 min	10–12 min	10–12 min
Braised (whole)	1½–2 hr	1–1½ hr	1–1½ hr	¾–1 hr
(sliced)	¾–1 hr	½–¾ hr	¼–½ hr.	½–¾ hr
Market form	Available fresh or frozen			

12.3 BRAINS

Figure 12.3 Brains: (left to right) lamb, pork, veal, beef.

Considered a delicacy by gourmets, brains are soft, tender, and mild in flavor. The major difference between beef, veal, pork, and lamb brains is their size. Beef brain is the largest, followed by veal, then pork and lamb. Brains are popular scrambled with eggs or prepared in a cream sauce for breakfast or lunch. There are also numerous recipes showing brains prepared *au gratin* (topped with grated cheese and baked), browned in butter, fried in batter, or jellied.

TABLE 12.4 FACTS ON BRAINS

	Beef	Pork	Veal	Lamb
Average weight	¾ lb	¼ lb	½ lb	¼ lb
Preparation time*				
Broiled or sautéed	10–15 min	10–15 min	10–15 min	10–15 min
Braised	20–25 min	20–25 min	20–25 min	20–25 min
Market form		Fresh, frozen, pickled		

*Brains must be parboiled in acidulated water prior to cooking.

12.4 SWEETBREADS

Figure 12.4 Sweetbreads

Sweetbreads are the thymus glands, which are located in the neck and near the heart. The thymus gland near the throat is typically round in appearance; that near the heart is more elongated. Of the two types, the one found near the heart is preferred. Sweetbreads are only present in young animals—that is, in lamb and veal. Similar to brains in tenderness and texture, sweetbreads are also considered a delicacy and thus command a higher price than other variety meats. Lamb sweetbreads are often used as garnishes for meat-stuffed pastries *vol-au-vent* and *timbales* or can be prepared *au gratin*, fried in batter, or skewered. Calves' sweetbreads are preferred in most recipes and can be prepared in a variety of ways such as browned in white or brown stock, *au gratin*, *à l'ancienne* (with kidneys, truffles, and mushrooms), or *à la Milanaise* (with grated cheese and bread crumbs).

TABLE 12.5 FACTS ON SWEETBREADS

	Veal (Calves)	Lamb
Average weight	½ lb	¹/₈ lb
Preparation time*		
Broiled or sautéed	10–15 min	10–15 min
Braised	20–25 min	20–25 min
Market form	Available fresh or frozen	

*Sweetbreads must be parboiled in acidulated water prior to cooking.

12.5 TONGUE

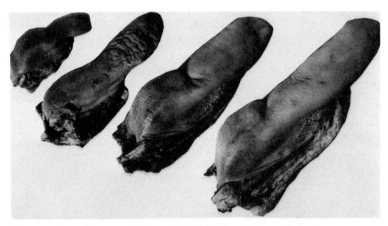

Figure 12.5   Tongue: (left to right) lamb, pork, veal, beef.

Being a well-used muscle, tongue is coarse in texture. It requires long, slow cooking in a liquid to become tender. The skin must be removed following cooking. Tongue can then be sliced and served hot or cold. Cured and smoked tongue is frequently sliced and used for making sandwiches. It also lends itself well to various other methods of preparation. Tongue can be prepared *au gratin*, with a medley of vegetables, with sauerkraut and bacon, or braised and then complemented with a number of sauces, such as mushroom, tomato, Madeira, or sweet and sour sauce.

TABLE 12.6 FACTS ON TONGUE

	Beef	Veal	Pork*	Lamb*
Average weight	3¾ lb	1½ lb	¾ lb	½ lb
Preparation time				
Simmered	3–4 hr	2–3 hr	1½ hr	1½ hr
Market form	Fresh, frozen, pickled, cured, smoked			

*Pork and lamb tongue are usually sold ready to serve. However, if purchased fresh, prepare as indicated.

12.6 HEART

Figure 12.6   *Heart: (left to right) beef, veal, pork, lamb.*

Heart is one of the more economical variety meats. It is also one of the toughest. Beef, veal, pork, and lamb hearts all require moist heat cooking as tenderization. Hearts are frequently stuffed with **forcemeat** (finely sieved chopped meat) or another suitable stuffing, braised, roasted, or baked in a casserole. Meat patties, meat loaves, and stews of other meats often include diced or ground heart for variety.

TABLE 12.7   **FACTS ON HEART**

	Beef	**Veal**	**Pork**	**Lamb**
Average weight	3–4 lb	¾–½ lb	½ lb	¼ lb
Preparation time				
Braised (whole)	3–4 hr	2½–3 hr	2½–3 hr	2½–3 hr
(sliced)	1½–2 hr	...	...	...
Market form		Fresh, frozen, pickled		

12.7 TRIPE

Tripe comes only from beef. Plain tripe is the muscle of the animal's first stomach; honeycomb is muscle of the second stomach. Honeycomb tripe is preferred. Prior to sale, the stomachs of cattle are washed, soaked in water containing a percentage of lime, scraped, and partially cooked. Tripe can be prepared in a variety of ways: fried in bread crumbs, grilled *à l'espagnole*, or as tripe *à la polonaise*, *à la portugaise*, *à la poulette*, or *à la provençale*, each of which has a different method of preparation and different ingredients to accompany the tripe.

TABLE 12.8 FACTS ON TRIPE

Source	Beef stomach
Average weight	1½ lb
Preparation time*	
Broiled or sautéed	10–15 min
Market form	Fresh, pickled, canned

*Tripe requires parboiling prior to cooking.

TABLE 12.9 ADDITIONAL VARIETY MEATS

Meat	Identification
Oxtail	Caudal appendage of ox's (beef) body. Generally used for oxtail soup or stew. Can also be stuffed, braised, or grilled.
Animelles	Testicles of male animals, particularly sheep or lamb. Often sautéed, fried in batter, or fricasseed.
Pig's ears	Pork offal used primarily for head cheese. Can also be stuffed, braised, or grilled.
Pig's feet	Pork offal, generally simmered or braised. Can also be pickled.
Pig's lungs	Pork offal used primarily for pork pâtés or stews.
Calf's ear	Veal offal, generally braised, grilled, or fried.
Calf's feet	Veal offal, generally simmered, fried, or deep fat fried.
Calf's head	Veal offal, generally simmered, stuffed, or fried.
Calf's lung	Veal offal, generally prepared in a ragout.

12.8 SUMMARY

Although variety meats account for only a small percentage of total meat consumption, they can be the basis for truly distinctive menu items. Restaurants have been known to build a reputation on their special preparation of sweetbreads, kidney, or liver. Offering well-prepared variety meats can enhance the menu, drawing new customers and adding profit.

CASE STUDY

It's Tuesday and the receiving clerk of the restaurant you manage has the day off and your head chef just called in sick. Luckily your relief chef can follow a standard recipe, and Tuesdays are normally slow.

Unbeknownst to you, an order placed by your purchasing agent with a new purveyor has arrived during the busiest time of the day. No one else was available, so the dishwasher accepted the shipment.

Later that evening you receive several complaints from customers. One guest says the calves' liver has an unusual strong flavor; another guest says the veal kidneys are very tough. You assure each guest that you will take care of the problem immediately.

You return to the kitchen and demand an explanation from the relief chef, who confesses that he doesn't know much about variety meats but assures you that he followed the standard recipe exactly the way it was written.

You check the refrigerator temperature; it is 34°F. You check the perpetual inventory card and find notices that a delivery of calves' liver and veal kidneys arrived today. Upon inspection, you notice that the calves' liver is slightly darker than usual and has three lobes about equal in size and that the veal kidneys are considerably darker in color, somewhat larger than normal, and have numerous lobes.

Your relief chef asks you if you know what caused the problem. Are the products spoiled? Were the right products accepted? (Consider individual characteristics of each variety meat.) How could this problem be avoided in the future?

13 | PROCESSED MEAT PRODUCTS

13.1 CURING

Curing is one of the oldest methods of preservation recorded in history. Curing can improve the flavor and palatability of meat and add variety to the diet. Ham and bacon are just two of the many cured products popular in the American diet. Supermarket shelves are lined with cured products to meet consumer demands, and almost any menu lists one or two cured items.

Curing, other than for home use, must be done commercially. Restaurants and stores are not permitted to cure on premises. Federal regulations require a federal inspector to examine all cured products for weights, internal temperatures, and ingredient levels before the finished product can be packaged and labeled for consumption.

Kinds of Meat Cured

Pork is the meat for which curing is most often used. Even under refrigeration, fresh pork spoils rapidly, so preservation is necessary if it is to keep. Curing is the primary method of pork preservation. The high fat content of pork makes it suitable for curing because the fat protects the lean from hardening during the curing process. The most common cuts of pork cured are hams, shoulders, bellies, and loins. Lighter weight pigs produce a more tender, lighter colored cured product. Hams weighing under 20 lb are usually preferred to heavier ones. Hams weighing over 20 lb are usually less tender and have a coarser texture and a darker colored lean. The same is true of bellies and pork shoulders as well. For cured and smoked Canadian bacon heavy boneless loins, weighing from 5 to 9 lb are usually required.[1]

Much less beef is cured than pork. Corned beef, which is made by curing brisket, round, or shoulder clod, and pastrami, made from plate, are the most common cured beef products. The top round or knuckle are also processed as dried beef. Poultry meat is often blended with other meats to make cured products such as sausage, frankfurters, and luncheon meats. "Turkey ham" and "turkey pastrami" have become popular substitutes for ham (pork) and pastrami (beef).

Very little veal or lamb is cured. The low level of production and the high cost of lamb and veal preclude wide use of these meats for cured products.

Ingredients

The basic ingredients for curing are salt (NaCl), sugar (sucrose), and sodium nitrite. In addition to these, ascorbates and phosphates and other preservatives are often added. Each of these ingredients fulfills a certain purpose in the curing process, and care must be exercised in adding and blending them. Certain manufacturers prepare commercial cure mixes in which the ingredients are carefully proportioned to meet federal regulations. Prepared mixes are extremely beneficial to smaller processors because they eliminate the need to carefully weigh out the different ingredients each time.

SALT   All cure mixtures are based on salt. Prior to commercial processing, salt was the only ingredient used for curing. By dehydration and osmosis, salt acts to reduce the amount of moisture present in meat, which in turn inhibits bacterial growth and spoilage. When used alone, salt produces a salty, harsh, dry product with an undesirable dark colored lean. To counter these effects of salt, sugar and nitrite or nitrate are commonly used in conjunction with salt. Certain fatty cuts of meat (clear plates, jowls, and fatbacks) are still often cured with salt alone, but these cuts are primarily for seasoning.[2] The amount of salt used per 100 lb of meat varies from 5 to 9 lb, the most common amount being 8 or 9 lb.

SUGAR AND OTHER SWEETENERS   Sugar performs several functions during the curing process. Although primarily added for flavor, sugar also counteracts the harshness of salt by retaining some of the moisture in the product. This in turn keeps the muscles soft and tender. Upon cooking, sugar causes the product to turn brown. The degree of browning (caramelization) depends partly on the type of sugar used. Sugar (sucrose) often causes the product to brown too much, producing an undesirable burnt flavor. Corn syrup, molasses, and other sugar substitutes are frequently used in place of sugar to reduce this browning effect or to add variety to the flavor. The amount of sugar used per 100 pounds of meat usually ranges from 2 to 4 pounds. Increased amounts of sugar can be added, but the high cost of sugar generally prohibits this practice.

NITRATE/NITRITE   Nitrite (sodium nitrite) is extremely important in the curing process. It is responsible for 1) fixing color, 2) preventing a warmed-over flavor by retarding the development of oxidative rancidity, 3) improving the flavor of cured meats, and 4) inhibiting the growth of poisonous microorganisms, particularly Clostridium botulinum.

During curing, the nitrite is initially reduced to nitric oxide in the meat, either by the natural chemical activity of post mortem muscle tissue or by the addition of reducing agents (ascorbates) to the curing mixture.[3] Nitric oxide then combines with the pigment myoglobin to form nitrosomyoglobin. At this point the lean appears to be dark red in color. Upon heating at temperatures of $130°-140°F$, nitrosomyoglobin is converted to nitrosohemochrome, producing the light pink color characteristic of cured meat. When the meat is packaged properly so that it is tightly sealed (by vacuum packaging), the nitrosohemochrome in the meat is a stable compound. If left exposed to oxygen or high-intensity light, however, the pigment changes to a gray-brown color. Nitrosohemochrome is also affected by certain metals. When aluminum, iron, or copper come in contact with the cured meat, peroxides are formed, a reaction that produces an undesirable green color. Sometimes when light strikes the surface of cured meats a fluorescent green color appears on the surface. This green fluorescence should not be mistaken for the color produced by metals; it does not affect the quality of the meat.[4]

Nitrite (NO_2) is the most controversial ingredient used in curing meats. When nitrite combines with secondary amines, a compound called nitrosamine is formed. Research has shown that when laboratory animals were fed large quantities of preformed nitrosamines some developed forms of cancer. Other tests have shown that practically all cured meat products are free of nitrosamines, however. Traces of nitrosopyrrolidine, a nitrosamine, have been found in some tested bacon samples but only when the bacon was fried past the crisp stage.[5] Nitrite is not only present in cured products. Vegetables such as beets and lettuce, human saliva, and water also contain quantities of nitrate (which break down to nitrite). Vegetables contain as much as 800–4,000 parts per million (ppm) of nitrate, and saliva has about 6–12 ppm.[6] At present, the advantages of using nitrite outweigh the disadvantages. Currently, a limit of 156 ppm of nitrite residual is allowed by the Food and Drug Administration in the finished product, but the limit may be made even smaller by more stringent regulations in the future. Bacon is already being limited to 120 ppm of nitrite residual and 550 ppm of ascorbate.

ASCORBATES Sodium ascorbate and sodium erythorbate (similar to ascorbate) are often added to the cure mixture. Ascorbate hastens the conversion of nitrite to nitric oxide and reduces the curing time. Ascorbate is also responsible for color stabilization during storage.

PHOSPHATES The primary function of phosphates is to increase the water-binding capacity of the muscle protein during curing. The addition of phosphates causes the product to retain more moisture, producing a more desirable, juicier product. Shrinkage is reduced and the yield or net weight of the product is consequently increased. Some evidence shows that phosphates may also aid in retarding oxidative rancidity and improving color stability. Sodium tripolyphosphate, sodium hexametaphosphate, sodium acid pyrophosphate, sodium pyrophosphate, monosodium phosphate, and disodium phosphate have been approved by the USDA for use in curing brines, but the amount is restricted to not more than 0.5 percent in the finished product.[7]

BAKING SODA   Occasionally, baking soda is added to the brine cure to keep the brine from souring. Only a small amount of baking soda is added, usually 2–3 oz per 100 lb of meat.

OTHER INGREDIENTS   Water is used in all but the dry cures as a transport medium, to distribute the curing agents throughout the meat, and as a means of preventing weight loss. Cured products are labeled according to the amount of moisture retained in the finished product (see section on labeling for further details). Sometimes spices are added to impart a particular flavor to the product.

Methods of Curing

Historically, meat was cured by salting alone. (Forms of nitrate present in the caves where the meat was cured explains why curing was so successful in those days.) Sometime later, it was discovered that this method could be improved by adding ingredients and immersing the product in a brine solution. Now meat can be cured a number of ways, including artery pumping, stitch pumping, and multiple injections. The latter two methods are the most common commercially.

Only meat from healthy animals must be used, of course. After slaughtering, the fresh, warm carcass is chilled at temperatures below 40°F. The meat should be prepared for curing as soon as the carcass has been completely chilled. Some sausages are produced using a hot boning method whereby, after slaughtering, the warm carcass is totally boned and the meat is chopped and processed. Hot boning is performed before rigor mortis sets in, and some evidence shows that the muscles retain more water using this procedure.

DRY CURING   There are two principal types of dry cures, using salt or salt with nitrate or nitrite and another one using the same ingredients with sugar added. The first type, dry salt cure, is used primarily on fatty cuts of meat (clear plate, jowl, and fatback). The end product is typically salty and has a hard surface. The other type, dry sugar cure, is used especially for dry-cured bacon and country-cured hams. The product is still a little salty, but the sugar eliminates some of the harsh taste of salt. The dry sugar cure method produces a less perishable product, but the process is slow and expensive.

The first step for preparing a dry cure is to mix the ingredients together and then rub them on the surface of the product. Pork bellies usually require only one rubbing before the product is placed on the shelf or in a box to cure. Larger cuts, such as the hams or pork shoulders, require several applications of the curing mixture. Depending on the consumer's preference, a number of different formulas can be used.

The time required for curing the product depends on the weight and thickness of the product. Generally, hams and bacons require at least 25–30 days for curing (2–3 days per pound) and are often cured for as long as 8 weeks. Country hams are usually cured 1 day per pound of ham, smoked at 80°F for up to 30 days, and then aged for 7 to 11 months. Bacon and loins, on the other hand, usually require only 2–3 weeks for curing (1–2 days per pound).

Regardless of the curing method used, the majority of the products are placed under refrigeration at 36°–40°F for the duration of the curing process. Higher temperatures accelerate the process, but they also tend to encourage more bacterial growth.

BRINE OR PICKLE CURING When water is added to the salt and nitrite, the combination is called a brine or pickle. When sugar is added to this mixture, the solution is called a sweet pickle. The strength of the solution varies with the amount of water added: the more water added, the less salty the solution. The salinity of the pickle is measured by an instrument called a salinometer (sometimes written salometer). A salinometer resembles a hydrometer. It is a ballasted glass vacuum tube graduated in degrees saline. To test the salinity of the brine, the solution is placed in a hydrometer cylinder or jar. The salinometer is then placed in the cylinder, and a reading is taken at the point where the salinometer meets the surface of the water. A reading of 100° would indicate a brine that was 100 percent saturated. Brine readings of 55° to 70° are more commonly used.

Once the ingredients are combined (using formulas similar to the ones used for dry curing) and the desired salinity of the brine is reached, the products are submerged in the pickle until the cure penetrates the meat. The curing time is usually less than the time needed for dry curing but more than the time needed for stitch pumping or multiple injections. A sweet pickle cure produces a more mild flavored product than dry curing, with a smaller percentage of shrinkage. The finished product is less dry and hard. Brine curing does not require as much labor as dry curing, but care must be taken to make certain that the brine does not sour while the product is still curing. Bacon can be cured by this method, although other methods of curing bacon are more common commercially. When boneless cuts of beef—brisket, plate, round, or chuck—are cured in a brine, the procedure is called corning. Most commercial establishments use the brine methods of curing for small meat items such as the tongue and corned beef.

Artery Pumping This method of curing is not as common today as it was in the past. Technological advances have provided processors with more efficient methods of curing. As the name suggests, the curing brine solution is pumped into the artery of the product being cured. Only cuts bearing arteries (hams and picnic shoulders) can be cured by this method. Special care is needed to avoid cutting off the artery when fabricating the wholesale cut. The artery must be left long when the carcass is fabricated so that the needle that will inject the brine can be easily inserted. The percentage of brine injected varies, ranging from 8–10 percent of the "green" (meaning fresh, uncured) weight of the product. The strength of the brine solution varies from 65° to 80° salinity; the less saline brine is used more often. Artery pumping is a slow process requiring skilled labor, and careful pumping is necessary to avoid rupturing the blood vessels. The product is also more perishable than a product dry cured.

Stitch Pumping Another method of curing is stitch pumping. Once again, a formula of dry ingredients is combined with water to produce a brine. The salinity of the solution may vary from 55° to 80° salinity. Typically a perforated needle, about 6 in. long, is used, connected to a power-driven pump resembling a gun, which is attached to a supply line immersed in the brine solution. The needle is inserted into the meat, and when the trigger is pulled, the brine enters the product. Usually three to six insertions are needed to inject sufficient brine. The amount of brine injected is usually 10 percent of the green weight, but it may be as high as 16 percent, depending on the desired salinity or intended cooking process. The needle should be inserted in several different places to assure uniform distribution of the brine. Some of the brine should be inserted close to the bone to prevent bone souring. (Bone souring occurs most often in slow-cured products like Smithfield ham.) The product is then covered in a brine until cured.

A hot pickling brine cure, 135°–140°F, 70° salinity, can also be injected into the product. Hams should not be kept in the "hot pickle" for more than 1 hour because this method accelerates the rate of curing considerably. Hot dry cures can also be injected by perforating the surface of the meat and applying the hot cure, but this method is good only for thin cuts. The temperature for dry cures is reduced to 48° to 50°F but a longer time, 3–5 days, is needed.[8]

Multiple Injection Of all the curing methods mentioned, this procedure is the fastest. Multiple needles are used instead of just one needle. The needles are mounted on a retractable holder so that they can be injected simultaneously into the product. When a needle strikes a bone, it stops but the other needles continue until injection of the brine is complete.

Production time and labor costs are considerably reduced with this method. The distribution of brine is excellent, since the needles are spaced close together and can inject the brine uniformly.

Labeling

All cured products are subject to federal inspection. Prior to curing, the products are weighed (usually by a lot) and the green, or fresh, weight of the product is recorded. After curing, the weight is recorded again; if the product is smoked, the weight is recorded once more. Before the product can be packaged and labeled, the inspector checks the weight of the finished product to determine how much moisture, if any, was retained. For example, if the finished product (say, a ham) weighs the same as the green weight or less, the product is labeled "Cured Ham." If the ham weighs up to but not exceeding 110 percent of the green weight, the product is labeled "Ham—Water Added." If the ham weighs more than 110 percent of the green weight, it must be labeled "Imitation Ham."

At temperatures of 137°F or greater, any trichinae organisms present in fresh pork are killed (see discussion of trichinosis in Chapter 6, Pork). Government regulations require most cured pork products to reach an internal temperature of 142°F during processing. Products labeled "Fully Cooked" must be cooked to an internal temperature of 155°F.

Table 13.1 is a list of cured and smoked pork products (see Tables 13.2–13.6 for sausages). Most products require further cooking unless labeled "Fully Cooked."

Since 1979, in response to consumers' fears of the dangers of nitrites, the meat industry has been permitted by the USDA to sell certain processed products without nitrites. The labels of nitrite-free products must bear the word "Uncured" as part of the product name and the statement "No nitrate or nitrite added" must also be printed on the label. If no alternative methods of preservation, such as canning or pickling have been used, the product must also be labeled "Not preserved–keep refrigerated below 40 degrees Fahrenheit at all times." Because of their high perishability, uncured products must be carefully refrigerated and stored like any other fresh meat product.

13.2 SMOKING

Smoking is another method of preserving meat and is often used along with curing. By inhibiting fat oxidization and bacterial growth, smoking extends the storage life of the finished product. Smoking also produces a more tender product. Tenderization is accomplished by meat enzymes, the activity of which increases because of the elevated temperature and humidity of the smokehouse.

The primary purpose of smoking is to enhance the flavor of cured meats. Most cured products are also smoked. Consumers enjoy the smoked flavor and are willing to pay the extra expense. As a result of smoking, the appearance and aroma of the product also improves. Heating stabilizes the cured meat color, while smoking gives meat an attractive glossy brown finish.

Although some beef is smoked, pork is the principal meat smoked. Poultry and fish can be smoked also. Salmon, herring, halibut, haddock, mackerel, sturgeon, eel, and carp are the most common types of fish smoked. Each type of smoked fish has its own unique flavor and would add variety to any meal.

Composition of Smoke

There are more than 200 compounds found in wood smoke. Four of these compounds play a significant role in smoking: phenols, carbonyls, organic acids, and alcohols. Both phenols and carbonyls are responsible for the smoky flavor found in smoked products. The phenols also act as an antioxidant and aid in preservation; the carbonyls give smoked products their characteristic color and aroma. Although organic acids have

TABLE 13.1   CURED AND SMOKED PORK PRODUCTS

Product	Identification
Bacon	A cut from the belly, cured, smoked, and partially cooked.
Calla (Callie) ham	A cut from the pork shoulder. Cured and smoked in the same manner as ham.
Canadian bacon	A cut from the loin. Cured, smoked, and cooked.
Cooked ham	Fully cooked ham, cured, may or may not be smoked. Usually sliced for sandwiches.
Country ham	Dry cured 30–40 days, slowly smoked and aged 7 to 11 months. Often named after the state in which it was produced; e.g., Virginia Ham.
Country-style ham	Has the same characteristics as country ham but is not produced in the country; e.g., Virginia-Style Ham.
Cured ham	Cured only. Labeled "ham" when cured weight equals green weight.
Ham—water added	Cured ham retaining up to 10% added weight due to absorption of curing solution.
Hickory-smoked ham	Cured ham, smoked over burning hickory chips in smokehouse.
Honey-glazed ham	Cured and smoked ham with a glaze of honey, sugar, and spices applied after smoking. Canned ham has glaze applied prior to processing.
Imitation ham	Cured ham retaining more than 10% added weight due to absorption of curing solution.
Jowl bacon	Cured, smoked, and partially cooked pork jowl.
Picnic	Cut from the front part of the pork shoulder. Cured in the same manner as ham.
Prosciutto	Dry-cured Italian hams manufactured from certified trichina-free hams. Requires several applications of dry cure mixture for a total of 45 days. Hams are soaked in 80° to 90°F water to soften skin, then smoked, cooled, rubbed with black and white pepper and aged 30 days at 70° to 75°F, relative humidity 65% to 75%. Eaten without cooking.
Salt pork	Salt-cured clear plates.
Scotch ham	Skinned, boned, fat removed, and cured (mild cure).
Smithfield ham	Processed in Smithfield, Virginia. Hand-rubbed with saltpeter and fine salt, cured, cool-smoked (80–85°F) for 7–10 days with hickory wood and apple wood sawdust, rubbed with black pepper, and then aged 7–18 months.
Smoked butt	Boston butt cut from the pork shoulder. Cured and smoked in the same manner as ham.
Smoked ham	Cured and smoked over burning hardwood in smokehouse.
Smoked Picnic	Cut from the front of the pork shoulder. Cured and smoked in the same manner as ham.
Westphalian ham	Rubbed with salt and saltpeter and cured 2 weeks, then placed in 90° pickle and cured 2 weeks, removed from brine and stored in cool dry cellar to ripen for 1 month, soaked in water for 12 hours, and smoked with juniper twigs and berries over beechwood fire for 1 week.

Source: Lawrence Drake, ed. *The 1951 Meat Manual* (New York: Lebhar-Friedman Publications, 1951), pp. 152–155.

little effect on the flavor of smoked meats, they do play an important part in the coagulation of surface proteins. Skinless frankfurters and sausages rely on organic acids for the formation of their outside coverings.[9] The last of the four, alcohols, acts as a carrier for the other volatile compounds.

TYPES OF SMOKE Hardwoods are the best fuel for smoking. Hickory, apple, elm, cherry, and maplewood yield the most desirable smoke. The type of hardwood chosen depends on the price and availability of the wood and the flavor desired in the product. Hickory-smoked products appeal to most consumers and are generally in great demand. Hickory wood and apple wood sawdust are used to produce Smithfield hams. Westphalian ham is smoked with juniper twigs and berries over a beechwood fire. Most commercial smoking operations burn hardwood sawdust rather than pieces of wood because the sawdust is easier to utilize and produces a greater volume of smoke.

Softwoods such as pine are avoided as a source of fuel because they tend to impart an undesirable resinous flavor to meat. "Liquid smoke," a seasoning agent made from hardwoods and a small percentage of softwoods, produces excellent results, however. This is because the tar substances common to softwoods are filtered out before the softwoods are used. Using liquid smoke captures the flavor of natural smoke but at a lower cost. The amount of smoke flavoring can be controlled and there is no need to install a smoke generator. Commercially prepared liquid smoke solutions are usually diluted with water or acetic acid. The solution is sprayed over the product prior to cooking. Liquid smoke can also be applied by dipping the product or by adding the smoke flavoring to the curing formula, but spraying is most common.

SMOKEHOUSES AND SMOKING Smokehouses are constructed of any material that will keep smoke in and withstand the heat and humidity generated within, either concrete, brick, stone, or stainless steel. The size and construction depend on the processor's needs. Ease of maintenance is considered in making the final selection. The smoke is generated outside the smokehouse by controlling the combustion of moist sawdust or hardwood chips or by the friction of hardwood logs or boards that rub against a tempered steel disk spinning at high speed. The volume of air can be regulated by opening or closing dampers or by using a controlled fan that recirculates the air uniformly throughout the smokehouse. The smokehouse also has a temperature-control device and a humidity control device.

The exact temperature and relative humidity chosen depend on what product is to be smoked. The time required for smoking depends upon the smoke concentration, temperature, and size of the cuts of meat. The time also depends on the desired color of the finished product. The longer the smoking, the darker the color of the ham, ranging from light straw-color to mahogany color. The flavor also becomes more pronounced with longer smoking times.

Cooking is often carried out simultaneously with smoking, but there are exceptions. Bacon is cured and smoked only until the desired color is attained. It is not really cooked until it is to be eaten. Hams, picnics, and bacons are usually placed on hooks or racks in the smokehouse. Care is needed to prevent products touching each other on the racks, because where surfaces touch smoke is prevented from penetrating uniformly. Often hams and picnics are placed in cloth stockinettes to help retain their shape.

Once the product is ready for smoking and the desired temperature and relative humidity are set, the product remains in the smokehouse until the desired flavor, color, and internal temperature are reached. After smoking, a federal inspector from the U.S. Department of Agriculture checks the internal temperature and the lot weight to make sure that the product meets federal regulations. The meat is then tempered to room temperature (gradually cooled to room temperature) and finally placed under refrigeration, ready to be sold. Smoked meat or fish can be stored under refrigeration for approximately two to three weeks.

13.3 SAUSAGES

Sausage making started centuries ago and sausage has been popular ever since. Frankfurters, fresh pork sausages, bologna, and salami are just a few of the different sausages purchased and consumed daily. The selection of sausages available today is very large. Sausages range from mild to spicy in flavor and from soft to hard in texture. Prices also vary. The semidry or dry sausages like salami generally command higher prices in the market place compared to fresh pork sausages, cooked and/or smoked sausages or meat loaves.

Ingredients

There are many different recipes for each of the different types of sausage on the market. Manufacturers often have special formulas for certain sausage types that may vary the percentages of ingredients or use additional ingredients. This is not to say that they have carte blanche when it comes to designing their own formulas, however. The U.S. Department of Agriculture strictly regulates the percentages of ingredients allowed in sausages. Especially important to the consumer are restrictions on the percentage of extenders and moisture. For example, semidry sausages are only allowed a maximum of 3.7 parts moisture for every 1 part protein in the finished product. Pepperoni, a dry sausage, must have a 1.6:1

moisture-to-protein ratio in the finished product. Prior to processing, all manufacturers must have their labels approved by the USDA. The manufacturer must send a copy of the formula and the processing procedure to the USDA for approval. Only when the product meets federal regulations can a label, bearing a specific name, be applied to the finished product. Most sausages use one or more of the following ingredients: meat, curing ingredients, fats, binders or extenders, seasonings.

MEAT Meat is the basis of all sausages, of course, but the type and percentage varies with the type of sausage. Most sausages are made of pork or beef, but veal and other kinds of meat may be used. Better quality sausages are made from better quality cuts of meat. Trimmings from primal cuts are used most often. The ratio of fat to lean differs depending on the type of sausage. Many of the less desirable cuts of meat—cheeks, hearts, livers, tongues, and snouts—are also used in various sausages, but they are self limiting because of their structure and other characteristics of their meat.

CURING INGREDIENTS Aside from fresh sausage, practically all sausages use the basic curing ingredients (salt, sugar, ascorbate, and nitrite) for processing. Once again, the percentage of nitrite used is stipulated by the federal government and depends on the type of sausage being produced.

FATS Fats are primarily to improve the flavor and sliceability of sausages made from an emulsion—that is, a mixture of meat, fat, and water stabilized by protein. Using more fats also decreases the cost of manufacturing the sausage. For a stable emulsion, sufficient amounts of protein are included in the formula to encapsulate the fat and hold it suspended in the product. The amount of fat added depends on the type of sausage. Dry sausages are processed using lean cuts of meat with a low fat ratio. Higher levels of fat tend to produce small holes in the finished product. Emulsion sausages are almost always cooked before they are sold.

BINDERS OR EXTENDERS Binders or extenders are used in cooked and smoked sausages as well as in commercially prepared meat loaves. They play an important role in fat encapsulation, in addition to improving the water binding capacity of the product, improving the flavor, texture, and slicing, and reducing the shrinkage during cooking. Dried milk, soy flour, soy grits, or soy isolates and soy concentrates—all rich in protein—also increase the protein content of the finished product. Their percentages are carefully regulated by the government. Sausages with a specific name are limited to no more than 3.5 percent soy flour, grits, concentrates, dried milk, cereals, and/or starch or 2 percent soy isolates in the total sausage product weight. Nonspecific products, meat loaves, for example, are not restricted to any special amount.

SEASONINGS All sausages are seasoned to some extent. The seasoning ingredient chosen is often the ingredient that makes a particular sausage unique. The most common spices used are pepper, sage, paprika, nutmeg, mustard, fennel seed, dill seed, coriander seed, cayenne pepper, cardamon seed, and ginger. Celery, garlic, and onion salt are also well-used seasonings. In addition, meat loaves are commonly processed with one or more of the following: olives, pimentos, pickles, cheese, and onions.

Processing

Grinding, chopping, and mixing are common to all sausage making. Some sausages are cured, smoked, and/or cooked; others are prepared fresh. Some are packaged loose; others are stuffed into casings, linked, or tied. Certain processed meats, meat loaves for example, are usually pressed into molds varying in size and shapes. Consequently, the exact procedure followed will depend on the type of sausage manufactured.

All sausages are ground. Grinding reduces pieces of meat of irregular size and shape to a more uniform size. The meat can be ground either coarse or fine. Some formulas call for both coarsely ground meat and finely ground meat, particularly when different kinds of meat, such as pork and beef (generally ground separately) are combined.

Mixing and chopping are often carried out simultaneously. The meat is chopped to the desired texture and the ingredients are uniformly mixed forming an emulsion. If the sausage is to be cured, the curing ingredients are added to the meat at this stage. Salt, spices, and any other additional ingredients are also added. The ingredients are added in a series of small steps, starting with one type of meat, then the dry ingredients mixed together, then shaved ice or ice water, and then any other type of meat, if more than one is used.

Stuffing follows mixing and chopping. Not all sausages are stuffed into casing, but most are. The type of casing depends on the type of sausage. There are two major types of casing: natural and artificial. Natural casings are made from the intestines, colons, or bungs of hogs, sheep, and cattle. They are used primarily for semidry and dry sausages. Natural casings are edible and shrink with the sausage when it is cooked. Artificial casings can be further broken down into two categories: cellulose, which is a byproduct of the cotton industry, and regenerated collagen, which comes from the collagen found on the inside or skin of beef hides. Most casings today are of cellulose. They are inedible and must be removed prior to eating. They may be water-permeable or nonpermeable, fibrous or nonfibrous, and will not shrink with the sausage. Most significant to the manufacturer, cellulose casings are inexpensive. Regenerated collagen casings are similar to natural casings, but are less expensive. Like natural casings, they are edible and shrink with the sausage. Care must be taken when stuffing the emulsion into the casings. Enough pressure must be exerted to prevent air pockets from forming. Once the casings are stuffed, they are tied or linked. Small sausages are generally linked, large sausages tied.

Next the sausages are hung on "sausage trees," evenly spaced. Fresh sausages are chilled to 32°F and then packaged for shipment. If they are to be sold as smoked cooked sausages or as unsmoked cooked sausages, they are cooked (and smoked) in the smokehouse. Next the cooked sausages are often showered with cold water to reduce the internal meat temperature and to set the protein skin, tempered to room temperature, and then chilled before packaging. Dry and semidry sausages are placed in a "green room" to ripen, be smoked (some are not smoked), and dry. The temperature and time for each step depends on the type of sausage being processed. Most sausages are held under refrigeration until they are ready to use.

Classification of Sausage Products

Sausages are classified as fresh, cooked, smoked and cooked, semidry and dry. Most sausages produced in the United States are types that originated in Europe. Many are named for their city of origin: bologna for Bologna, Italy; frankfurters for Frankfurt, Germany; wieners for Vienna, Austria; Genoa salami for Genoa, Italy; and Braunschweiger for Braunschweig, Germany. Fresh and cooked sausage originated in Northern Europe where the colder climate could preserve them, whereas dry and semidry sausages originated in the Mediterranean countries, where the warmer climate necessitated preservation by drying and smoking. Although a wide variety of sausages is available on the market, only the most common are presented in this chapter. Examples of each class of sausage are listed in the tables in the following sections.

Fresh Sausages Fresh sausage is made from fresh, uncured cuts of meat. Pork trimmings are the principal ingredient used in fresh sausage. Certain varieties of fresh sausage may also be prepared using cuts of beef, veal, and variety meats. Water or ice may be used to facilitate mixing or chopping but the amount used may not exceed 3% of the total ingredients used. Fresh sausage is highly perishable and must be stored under refrigeration. Cooking is required before eating. The following is a list of the different types of fresh sausage.

TABLE 13.2   FRESH SAUSAGE

Example	Contents and Market Form
Fresh pork sausage	Prepared using pork trimmings. Sage is the principal seasoning. Available in different size links, patties, and bulk.
Fresh country-style pork sausage	More coarsely ground than fresh pork sausage and may contain 10–20% ground beef. Casings or bulk.
Fresh breakfast-style pork sausage	Ground pork seasoned with salt, pepper, and sage. Links.
Italian pork sausage	Pork trimmings. The "hot" variety is seasoned primarily with red pepper, coriander, and fennel seeds. The "sweet" or "mild" variety is seasoned primarily with fennel seeds and Spanish paprika. Wide casings and linked.
Bratwurst	The German name is used for this pork sausage that may also contain cuts of veal. Slightly flavored with the juice or rind of a lemon. Linked casings, 4 in. long and about 1½ in. in diameter.
Fresh sausage or sausage meat	Pork is the principal ingredient but may also contain cuts of beef and variety meats. Links, patties, or bulk.
Bockwurst (white sausage)	A combination of veal, pork, milk, and eggs, seasoned with parsley and chives. Links about 4 in. long.
Fresh Thuringer-style sausage	Pork is the principal ingredient. Sold 8–10 in. long and about 1½ in. in diameter.

Sources: Lawrence Drake, ed. *The 1951 Meat Manual* (New York: Lebhar-Friedman Publications, 1951), pp. 152–155. Stephan L. Komarik, Donald K. Tressler, and Lucy Long. *Food Products Formulary: Volume 1 Meats, Poultry, Fish, Shellfish* (Westport, Conn.: Avi, 1974), pp. 26–87. W. E. Kramlich, A. M. Pearson, and F. W. Tauber. *Processed Meats* (Westport, Conn.: Avi, 1973), pp. 182–219. John R. Romans and P. Thomas Ziegler. *The Meat We Eat*, 10th ed. (Danville, Ill.: Interstate, 1974), pp. 531–557. The same sources were used for Tables 13.3, 13.4, and 13.5.

Smoked and Cooked Sausages   Smoked and cooked sausage is the largest selling group of sausages. Some are cooked only. Cooked, smoked sausages may use water or ice in the formula to facilitate chopping and mixing, or to dissolve the curing ingredients, but no more than 10 percent of the added water may remain in the finished product weight. Although cooked and smoked sausages are less perishable than fresh sausages, they must still be kept under refrigeration. The frankfurter is the best selling item in the group.

TABLE 13.3 SMOKED AND COOKED SAUSAGES

Example	Contents and Market Form
Frankfurter	An emulsion-type sausage. There are three major types of frankfurters: beef, meat, and frankfurters made without nitrites. Frankfurters prepared from only beef and no other source of meat can no longer be labeled "all beef" because they contain other ingredients (corn syrup, salt, pepper, nutmeg, sodium erythorbate and sodium nitrite) in addition to beef. Kosher-style frankfurters are made only with beef and beef fat, no other meat. High-quality "meat" frankfurters are prepared using 60% beef and 40% pork. Lesser quality meat frankfurters use a lower percentage of beef and often use pork jowls instead of pork trimmings. Frankfurters prepared without nitrites are highly perishable. Frankfurters are cured (except the no-nitrite type), spiced, stuffed into casings, linked, smoked, and cooked. Frankfurters can be braided into groups of links (weiner style) or twisted into links (Vienna style). Available from ¾ in. in diameter and from 4 to 5½ in. long up to 1 ft long, or bite size (cocktail type).
Knockwurst	An emulsion-type sausage similar to the frankfurter, smoked and cooked. Links about 2½ oz. each.
Bologna	Emulsion-type sausage. The quality varies according to the cuts of meat used and whether or not extenders have been added. Higher quality bologna is prepared using a mixture of beef, pork, and seasonings, cured, smoked, and cooked. Available in ring form 1½ in. in diameter or in long form, 18–20 in. in length and 2½–4 in. in diameter.
Braunschweiger (liver sausage)	Emulsion-type sausage prepared from pork livers, jowls, and seasonings, cured, smoked, and cooked. Available in casings, 2¾–3 in. in diameter.

Semidry Sausages Semidry sausages, some called summer sausages, are characteristically drier than fresh and smoked sausage. About 50–55 percent of the moisture is removed as a result of the curing, smoking, and drying process. Subsequently semidry sausages can be held under refrigeration for a longer period of time in comparison to fresh sausage. Starter cultures of bacteria are often added to the semidry sausage mixture of ground meats, spices, and curing ingredients to start the fermentation process and reduce the processing time.

TABLE 13.4 SEMIDRY SAUSAGES

Example	Contents and Market Form
Cooked Salami	Prepared from pieces of beef and pork seasoned with garlic, then ground, cured, cooked, smoked, and dried. Most salami found on the market is the dry sausage type. Available in casings, each salami weighs 5–6 lb on the average.
Thuringer	From cuts of beef and pork trimmings seasoned primarily with ground black pepper and mustard, then ground, cured, smoked, and dried. Cooking occurs simultaneously with smoking until any trichinae present are killed. Available in casings; 2¼ lb – 7 lb.
Cervelat	From beef and pork trimmings, although sometimes tripe, beef cheeks, and pork stomachs are used. Seasoned primarily with ground or whole black pepper. Ground, cured, dried, and smoked. Cooking occurs simultaneously with smoking only until any trichinae present are killed. Semisoft cervelat is cooked to an internal temperature of 152°F. Available in casings.

Dry Sausages Dry sausages generally command a higher price in the marketplace because more skilled labor and time is needed to prepare this type of sausage. Certain varieties, e.g., dry salami, require as much as 90 days for drying. After drying, the sausage is ready-to-eat. No cooking is required during processing or prior to eating. About 65% or more of the moisture is removed as a result of the curing, ripening, smoking and drying process. Dry sausages have an excellent storage life and should be stored in a cool, well-ventilated room or under refrigeration. The following is a list of dry sausages.

TABLE 13.5 DRY SAUSAGES

Example	Contents and Market Form
Salami	The most common type of dry sausage on the market comes in a variety of different types (B.C. Salami, Genoa, Milano, Kosher). Salami is generally prepared from a combination of beef and pork (Kosher salami uses beef only), seasoned with pepper and garlic powder, cured, stuffed, ripened, wrapped with twine (varies with the type), occasionally smoked, and then dried for a long time (60–90 days). Size available depends on type. Milano, for example, comes 3½ in. in diameter, 25 in. in length.
Pepperoni	Prepared using pork trimmings, beef chucks, and sometimes pork hearts and cheeks. Seasoned primarily with pepper, cured, ripened, occasionally smoked, and then dried for approximately 3 weeks. Available in linked pairs about 10–12 in. long and 1¼ in. in diameter.
Cervelat	Also available as a semidry sausage. Comes in a variety of different types (Farmer, Goettinger, Goteborg, Gothaer, Holsteiner). The percentage of pork and beef, type of grind, seasonings, processing procedure, and size of sausage all vary according to the type.
Chorizos	Prepared primarily using pork trimmings, although certain recipes call for a mixture of beef trimmings, cheeks, and hearts. The meat for chorizos is highly seasoned with pepper, then cured, stuffed, ripened, smoked and dried for about 2 weeks. Available in links about 3–4 in. in length.

13.4 SUMMARY

Cured and smoked products are successful selling items. For years delicatessens have displayed different varieties of sausages and hams in their windows and display cases and for years their sales have increased. By their mere presence, sausages and smoked products sell themselves. One needs only smell their aroma for the gastric juices to start flowing. At least one well-known restaurant chain has built its fortune on a sausage theme. The old-fashioned delicatessen atmosphere can be brought to the modern restaurant which can profit too from the popularity of cured meats.

CASE STUDY

SAM's packing house has called your purchasing agent with a special offer. He informed the purchasing agent that he has a surplus of bacon, cooked ham, fresh breakfast-style pork sausage, frankfurters, and Genoa salami. If you take 150 lb of each today, he will give your restaurant a 10 percent discount.

According to your chef, the restaurant uses 35 lb of bacon per week, 40 lb of cooked ham, 20 lb of fresh breakfast-style pork sausage, 35 lb of frankfurters, and 10 lb of salami.

You check your inventory onhand and realize you have only 5 lb each of the bacon, sausage, and frankfurters, 10 lb of ham, and 2 lb of salami left in the refrigerator. Your freezer is full, but you have enough space in your walk-in cooler to store the bulk purchase.

What are your recommendations? Would your decision change if enough freezer space became available in five days? (Consider optimum storage time of cured meats and bulk purchasing factors.)

REFERENCES

1. W.E. Kramlich, A.M. Pearson, and F.W. Tauber, *Processed Meats* (Westport, Conn.: Avi, 1973), pp. 112–116.
2. Kramlich, Pearson, and Tauber, p. 41.
3. J.C. Forrest, Elton D. Aberle, Harold B. Hedrick, Max D. Judge, and Robert A. Merkel. *Principles of Meat Science* (San Francisco: W.H. Freeman, 1975), p. 197.
4. Robert E. Rust, and Dennis G. Olson, *Meat Curing Principles and Modern Practice* (Kansas City, Mo.: Koch Supplies, 1973), p. 5
5. American Meat Institute, *Nitrite* (Washington, D.C., AMI, January 1978), p. 3.
6. *The National Provisioner*, p. 36 (August 28, 1976).
7. Rust and Olson, p. 14.
8. Kramlich, Pearson and Tauber, p. 59.
9. Kramlich, Pearson and Tauber pp. 63–66.

14 | PURCHASING

The next several chapters deal with the flow of meat, fish and poultry through a commercial food service facility. The technical elements developed in the preceding chapters are the basis for the managerial skills necessary to control purchasing, receiving, storage, preparation, and cost analysis within a food service facility. If management control of meat, fish, and poultry within any one of these subsystems is weak, the entire system and ultimately the profitability of the operation suffer.

14.1 THE ESSENTIALS OF PURCHASING

In purchasing, the first step is to define the needs of the food service operation. One way to do this is to conduct a **market survey** to determine the mix of guests (the potential mix if the operation is new). A menu can then be designed to fit the market profile. As the menu is being developed, certain tests should be performed on the entrees to determine their suitability. (Grade, tenderness, and flavor should be tested to determine palatability, and yield should be assessed to determine the cost and decide what to charge the customers for each item.

Communication Among Purchasing Agent, Suppliers, and Receiving Department

The next important step is to communicate the purchasing needs to the various purveyors by developing a set of **purchasing specifications** for all of the menu items. A specification is a precise written statement describing in detail the exact product desired. Presenting well-defined written specifications to suppliers competing for the order allows the purchasing agent to compare price quotes from the different suppliers for the same items. The receiving clerk should also receive a copy of the purchasing specifications. The clerk can perform a better job if the product received can be checked against a detailed written order. If upon receipt the merchandise is properly inspected and checked against the purchasing specifications, there is some assurance of uniform quality and uniform cost. Uniformity in both quality and cost is extremely important for pricing menus and maintaining profitability.

14.2 ELEMENTS OF A PURCHASING SPECIFICATION

Meat specifications vary from one operation to the next, but all specifications should set forth the following: the name of the cut desired; the quality and yield grade (where applicable); the weight range of the product (allowing some tolerance); the fat and trim desired; in some cases, the designated number from the National Association of Meat Purveyors (NAMP) *Meat Buyer's Guide* (described in the next section); the condition of the product—fresh or frozen and possibly the internal temperature; the method of tenderization, if necessary; packaging material; and any other special considerations required. A product specification sheet like the example in Table 14.1 may be used for different meat items.

How to Develop Specifications

Specifications must be tailored to individual needs. Each operation has a different market, a different price structure, and a different method of preparing and presenting food. As a result, specifications from one operation are generally not appropriate for another. In the past ten years, with the advent of boxed beef, there has been increasing standardization of meat cuts. The National Association of Meat Purveyors' publication entitled *The Meat Buyer's Guide* is an excellent source of information on all standardized wholesale and portion cuts for beef, lamb, veal, and pork. It uses a numbering system whereby each of the major cuts is identified by a number. Since details on how to fabricate each numbered primal, wholesale, or portion cut are given in the book, it is convenient to use the NAMP number to specify cuts of meat when ordering.

Although the NAMP book aids a great deal in writing specifications, the NAMP number by itself is not enough. A well-written specification describes in detail all of the characteristics listed on the product specification sheet shown in Table 14.1. Variations from the

---

TABLE 14.1 PRODUCT SPECIFICATION SHEET

Menu item:
Product:
NAMP no.:
Grade:
Trim factor:

Weight range:

Method of tenderization:

Packaging:
Special considerations:

State of refrigeration:

standard cuts listed in the NAMP guide may be needed. For example, a restaurateur may want to purchase a #109 roast-ready rib with ribs only from the eighth to the twelfth section, rather than from the sixth to the twelfth. When writing a specification, however, one must keep in mind that any special treatment or handling of the product by the purveyor generally increases its purchase cost.

Reasonable levels of tolerance should be allowed for the weight range and product trim. Animals are not stamped out on an assembly line like cars, and there will always be some variation. It is important that there be limits on variation tolerated, of course, but one must be realistic when specifying the limits. The following is an example: A restaurant's agent orders 8-ounce top sirloin butt steaks and refuses to allow any variance from the 8-ounce portion. In order for a purveyor to meet this demand, steaks would have to be cut heavier than 8 ounces and trimmed back to the exact weight. This process is time-consuming and costly for the purveyor, and the cost is subsequently reflected in the price charged for the item. Specifying a ½-ounce tolerance could save the purchaser a substantial amount of money; customer satisfaction would not be affected if steaks ranged from 7½ ounces to 8½ ounces.

Why Food Service Managers Often Neglect Specifications

Many owners or managers of small food service operations believe that it is a waste of time to prepare written specifications. Their claim that they do not have the buying power to enforce their specifications may be an excuse for laziness, however. It is true that, because purveyors' profit is only pennies per pound, they require tonnage to be profitable. And it is also true that the small operation may not have the buying clout of a large chain or corporate operation. Nonetheless, its orders are still a very important part of the purveyor's volume figure.

Many operators also state that if they write well-defined specifications for their needs, the price on the merchandise purchased generally goes up. Although this often does happen, it is better to pay a slightly higher price in order to receive items of consistent quality and price over the long run than to pay a few pennies less per pound and purchase items that are unsuitable or inefficient for the operation. Specifications should be designed to enable the purchasing agent to procure items most suited for the operation at the most economical price. (This maxim will be repeated subsequently in sections on purchasing policy and procedures.)

Why Specifications Must Be Detailed and Precise

To suggest how much detail should be written into a purchasing specification, Tables 14.2–14.5 are hypothetical beef specifications written for different types of food service operation.

TABLE 14.2 PURCHASE SPECIFICATION FROM A FIRST CLASS HOTEL OR RESTAURANT

Menu item:	New York strip streak (12-oz boneless portion)
Product:	Strip loin
NAMP no.:	#180
Grade:	USDA Prime, yield 2
Trim factor:	Not to exceed ½ in. fat with smooth and even distribution.
Weight range:	10–12 lb
Method of tenderization:	Age 10 days
Packaging:	Cryovac
Special considerations:	Strip eyes must be at least 2½ in. wide but no more than 3½ in. wide.
State of refrigeration:	Fresh only, ship chilled at temperatures below 40°F.

FIRST CLASS HOTEL OR RESTAURANT An operation using a specification like Table 14.2 chooses to purchase wholesale cuts and fabricate portion-cut steaks on the premises. There are a number of reasons why its managers may choose to do so. First, they may wish to control the length of the aging period to guarantee a certain level of tenderness. Second, they may have a butcher on the premises to do the fabrication and they may also desire the flexibility in storage time afforded by the larger cut (the whole strip loin can be stored longer than portion-cut strip steaks).

Asking for a # 180 strip loin tells the purveyor that the strip loin must be boneless and that the tail on the strip loin must be cut 2 in. from the edge of the eye muscle at the loin end and 3 in. from the edge of the eye muscle at the rib end. This gives the purveyor the gross fabrication information needed; the fabrication requirement is further defined by the trim factor of ½ in. over the eye muscle and the weight range of 10–12 lb.

Specifying **weight range** is extremely important. The restaurant using the specification in Table 14.2 will be cutting 12-oz boneless steaks. If the weight of the strip loins from which the steaks will be fabricated were to fluctuate above 12 lb, it would be difficult to cut a good thick steak portion. (Strip loins in the heavier weight range have wider eye muscles and thus must be cut thinner to keep the portion to 12 oz.) Because the restaurant or hotel is concerned with consistency and with appearance of the menu item, it has further defined the characteristics of product to be purchased by requesting the purveyor to select strip loins that have eye muscles between 2½ and 3½ in. wide. This detail will undoubtedly increase the cost of obtaining the product, but for this type of operation, the slight increase in cost is minimal and uniformity is very important.

To guarantee the highest degree of palatability, a USDA Prime is specified. The grade specification could be further broken down into low or top Prime should the operation wish to do so. A yield number of 2 is listed, but it is not extremely critical in this case because the amount of exterior fat tolerated and the size of the strip eye muscles have already been specified. The most suitable method of tenderization for USDA Prime strip loins is aging. The operation has requested the purveyor to start the aging process by holding the strip loins for 10 days after slaughter. The hotel or restaurant will continue the aging process within its own facilities to assure the optimal level of tenderness for its customers. To reduce the possibility of drip loss and trim loss due to aging, the strip loin is ordered vacuum packaged in a plastic film and shipped chilled at temperatures below 40°F.

TABLE 14.3   PURCHASE SPECIFICATION FROM A FAMILY-TYPE RESTAURANT

Menu item:	New York strip steak (12-oz bone-in portion)
Product:	Bone-in strip steak
NAMP:	#1179
Grade:	USDA low Choice
Trim factor:	¼ in. fat cover, tail not to exceed 2 in.
Weight range:	12-oz portion with ½-oz tolerance
Method of tenderization:	Age whole strips 1 week prior to cutting. Also run whole strip through pinning device before steaks are cut.
Packaging:	Individually bivac—12 portions/box
Special considerations:	Steaks must be at least ¾ in. thick. Steaks with vein on both sides are not acceptable.
State of refrigeration:	Frozen, temperature must be maintained below 0°F.

FAMILY-TYPE RESTAURANT   Buying portion-cut steaks in 12-oz. bone-in portions (see Table 14.3) may be an attempt to give the appearance of a larger portion while keeping the cost of the item down. To guarantee tenderness of the quality grade requested—USDA low Choice—a combination of aging and mechanical tenderization prior to cutting the individual steaks has been specified. Fat cover, length of tail, and thickness of each portion are specified for product consistency. To prevent customer dissatisfaction with tough loin end steaks containing connective tissue, "steaks with vein on both sides" are specified unacceptable in this order. Finally, since steaks that have been aged and mechanically tenderized do not keep very long in the fresh state, they are to be individually vacuum wrapped in plastic film, frozen, and shipped at temperatures below 0°F.

FAST FOOD STEAK HOUSE   A fast food restaurant tries to produce acceptable-quality steak at very reasonable cost. The purchaser using the specification shown in Table 14.4 is ordering portion-cut, 14-oz strip steaks. This portion is large compared to steaks described in the previous specifications (Tables 14.2 and 14.3), but this steak has quite a bit more waste remaining on it, with a heavier fat cover, more tail, and the bone

TABLE 14.4 PURCHASE SPECIFICATION FROM A FAST FOOD STEAK HOUSE

Menu item:	Charbroiled strip loin
Product:	Bone-in strip steak
NAMP no.:	#1177
Grade:	USDA Commercial (cows only)
Trim factor:	Fat not to exceed ¾ in. Tail not to exceed 3 in.
Weight range:	14 oz with 1-oz tolerance. Thickness ¾ in. minimum.
Method of tenderization:	Steaks individually dipped in liquid tenderizer.
Packaging:	Multivac—24 portions/box
State of refrigeration:	Frozen, temperature must be maintained below 0°F.

left in. The tolerance allowed is up to 1 oz, which means a portion ranging in weight from 13 to 15 oz would be acceptable. The quality level is lower—USDA Commercial—and cows only are specified to prevent bull meat from being used. (Bull meat has a rather strong and objectionable odor as well as an unusually dark color, characteristics that make it undesirable for fresh meat consumption.) Because of the lack of tenderness in this grade range, the individual steaks will be chemically treated by the purveyor, before being packaged, frozen, and shipped at 0°F or below.

AIRLINE FEEDING The steak specified in Table 14.5 is a specialty item; a convection oven on the airplane would be used to bring the steak to the degree of doneness requested by the passenger. The airline is requiring minimal waste on the 10-oz portions. It is also ensuring a good level of palatability by specifying USDA top Choice beef to be aged for 2 weeks prior to fabrication of the steaks. To ensure appealing appearance, the steaks are seared for color development, but they are kept below 40°C internal temperature so that passengers requesting steaks rare to medium can be accommodated.

TABLE 14.5 PURCHASE SPECIFICATION FOR AIRLINE FEEDING, FIRST CLASS

Menu item:	New York strip steak
Product:	Preseared boneless strip steak
NAMP no.:	#1180 B
Grade:	USDA top Choice
Trim factor:	¼ in. fat cover, ½ in. tail
Weight range:	10 oz seared weight, ½-oz tolerance. Minimum steak thickness ¾ in.
Method of tenderization:	These steaks shall be cut from strip loins that have been aged for 2 weeks.
Packaging:	Steaks are individually packaged in special airline foil containers.
Special considerations:	Each steak will be seared on a charbroiled broiler until ideal color development occurs. Internal temperature of the steaks must not exceed 40°C (104°F) during the searing process. Steaks will then be frozen by liquid nitrogen to −5°F.
State of refrigeration:	Frozen, temperature must be maintained below −5°F.

The example beef purchase specifications shown in Tables 14.2–14.5 show the detail that should be written into each specification. This is important whether the restaurant specializes in $25.00 à la carte entrees or $2.50 complete dinners. One can readily see that it would be a mistake for the fast food steak house to attempt to use the specification written for the first class hotel or restaurant. Although the structure of the specification is the same, the details of the requirements differ as greatly as the market and prices of these two different types of food service operation.

Pork, lamb, and veal specifications are written using the same format as for beef. The following examples of other product specifications are given to provide the reader with a model which can be used for development of poultry, fish, and shellfish specs.

TABLE 14.6   FISH AND SHELLFISH PURCHASE SPECIFICATIONS

Menu item:	Shrimp Scampi
Product:	Brown shrimp
Count:	Under 10
Market form:	Green headless
Packaging:	5-lb box
Special consideration:	Shrimp must be properly glazed and show no signs of deterioration.
State of refrigeration:	Shrimp must be frozen and maintained at temperatures below 0°F.
Menu item:	Broiled rainbow trout
Product:	Rainbow trout
Weight range:	1 lb with a 1-oz tolerance
Market form:	Drawn
Special considerations:	Fish must smell fresh, be bright in color, with firm flesh and bulging eyes.
State of refrigeration:	28°F packed on crushed ice
Menu Item:	Fillet of sole
Kind:	Dover
Weight range:	8 oz with a ½-oz tolerance
Market form:	Fillet
Packaging:	Individually wrapped in moisture-proof plastic wrap
Special considerations:	Fish must be odorless, completely frozen, and show no signs of frost or deterioration.
State of refrigeration:	Individually quick frozen (IQF), temperatures maintained at below 0°F
Menu item:	Clams on the half shell
Kind:	Cherrystone
Market form:	Alive in shell
Special considerations:	Shells that have opened must remain closed when pressed; meat must be plump and creamy colored.
State of refrigeration:	Fresh, 34°–36°F

TABLE 14.7 POULTRY PURCHASE SPECIFICATIONS

Menu item:	Boneless stuffed squab
Kind:	Chicken
Type:	Fresh
Class:	Broiler-fryer
Size:	2–2¼ lb
Style:	Ready to cook
Grade:	A
Special considerations:	Birds must be fresh, well fleshed, free from defects, free from pinfeathers, and show no signs of deterioration.
Menu item:	Stuffed Rock Cornish game hen
Kind:	Chicken
Type:	Frozen
Class:	Rock Cornish game hen
Size:	1½ lb with a 1-oz tolerance
Style:	Ready to cook
Grade:	A
Special considerations:	Birds must be frozen solid and show no signs of deterioration or freezer burns.
State of refrigeration:	Frozen, temperatures maintained at below 0°F

14.3 SELECTING SOURCES OF SUPPLY

Once specifications have been developed, the next major step in purchasing is selecting the sources of supply. Information on suppliers can be obtained from various sources, including the Yellow Pages, trade publications, and purchasing agent associations. Having found a source of supply, check the integrity and reputation of the supplier immediately. This can be done through agencies such as Dun and Bradstreet or Standard and Poor. Call some of the purveyor's customers and ask for their opinion.

Number of Suppliers Needed

At least two or three purveyors should be used by most food operations. Often a restaurant owner states that he or she has done business exclusively for a certain number of years with one purveyor, having found that purveyor to be the most honest and reliable and most economical. How can one possibly judge a purveyor's honesty, reliability, and prices unless constant comparisons are made with other purveyors? Aside from the importance of healthy competition, purveyors and their salesman can be good information sources about supplies and price trends. Limiting the food service operation to one source of information can be dangerous and costly. But, on the other hand, so can having too many suppliers. Not only is using too many suppliers unprofitable for the purchaser, increasing costs of ordering and accounting, but it may be unprofitable for the supplier as well. A supplier may refuse to service an account because its purchase orders are too small (if the purchaser is dividing its orders among too many firms).

Evaluating Potential Sources of Supply

A meeting with the plant manager is the most effective way to judge a potential source of supply. Few food service operators take the time to perform this very valuable step in selecting purveyors. From a brief tour of the plant, many facts can be ascertained about the purveyor. The skill level of the employees in the plant can be evaluated by watching the fabrication techniques. The sanitation level throughout the plant can also be judged; sanitation affects the keeping quality of the product purchased from the purveyor. One can also learn if the purveyor carries a variety of grades and products. At the same time, it is important to learn what services the purveyor may provide. Is the plant equipped with modern equipment for mechanical tenderization of the product? Are the freezers and walk-in coolers adequate in size and operating conditions to hold products for extended periods of time? Does the purveyor's staff have the technical expertise necessary to age the product properly? All of these factors should be considered before selecting a firm as a source of supply.

From discussions with the plant manager, one can determine how informative the supplier will be about price trends, new products, and market conditions and how cooperative the staff will be in meeting deadlines and making scheduled and emergency deliveries. A very important factor is the terms of payment. Are discounts given for rapid payment or purchases in quantity? Carefully evaluate the supplier's buying power. Does the supplier have the capacity to meet your needs? Is there depth and competence in the supply firm's management? Without it, the supplier may go out of business, terminating your supply. Investigate the labor relations at the plant. How often are there strikes? This will affect service.

SUMMARY   It would be simple to set up a score card like the one shown in Table 14.8 to rate the various suppliers being assessed by a potential purchaser. It is essential to take the time to tour the purveyor's plant and speak with top management. One should look at the following:

1. Size. Can the plant adequately supply your present and future needs?
2. Staff. Is the staff sufficiently competent and knowledgable about the product and about business to meet your requirements and specifications?
3. Integrity and reputation. Call customers for their opinion.
4. Labor relations. How many strikes have occurred in recent years?
5. Sanitation and working conditions of the plant.
6. Purchasing power.
7. Financial position. Get a Dun and Bradstreet report (DBR) on the firm.
8. Management expertise and depth.
9. Product mix.
10. Ancillary services. Does the supplier offer storage and tenderization?

TABLE 14.8 SCORE CARD FOR EVALUATING A POTENTIAL SUPPLIER

Product	Ideal Score, %*
Quality	20
Consistency	10
Mix	5
New product development	5
Service	
Delivery schedule	10
On-time delivery	5
Storage service,	
short and long term	10
Technical services (tenderization)	10
Market information	5
Price	
Competitive prices	10
Terms—length of time quote holds	5
Discounts	5
	100%

*The weighting of the percentages will vary depending on the requirements of the purchaser. A preferred supplier rates 85% or higher; an acceptable supplier rates above 75%; an unacceptable supplier rates below 75%.

The supplier will be rating you on the following:
1. Knowledge. Are you knowledgable about the products you are buying? Are there defined specifications? Are they current? Are they realistic? Are they used?
2. Financial acumen. Do you understand the price/value relations?
3. Credit. Do you pay on time? Or do you require extended credit?
4. Delivery schedule. Do you comply with the agreed delivery sequence or are many extra deliveries called for?
5. Receiving. Are your receiving procedures adequate and consistent?
6. Purchasing method. Do you prefer total bid, line-by-line bid, cost plus or formulated pricing, or contract?
7. Volume. What is your current and future volume?

The Letter of Intent

Once the survey stage has been completed and various purveyors have been rated and certain ones selected as sources of supply, the food service facility must set ground rules for maintaining good business practice. A letter of intent spelling out in detail the policy of the purchasing department should be sent to each purveyor likely to be used. The following letter is a sample:

XYZ Meat Packing Company
Steer Street
Packing House, USA

Gentlemen:

We are happy to inform you that you have been selected as a

potential source of supply for our restaurant facility. In order for both of our businesses to prosper, we must develop a good working relationship. This cover letter, attached to our specifications, is sent to you as a first step toward developing good business relations.

All meats—primals, wholesale cuts, portion cuts, variety meats, processed or preprepared items—shall be purchased through USDA inspected purveyors only. Beef, lamb, and veal shall be graded Prime or Choice by a USDA official and no other grades will be received.

All merchandise shall be in excellent condition when delivered and must conform to our specifications. All items shall meet the requirements of weight range, grade, condition, and quality at all times. Tolerances will be limited to adverse market conditions and only with our prior approval.

All products will be closely examined upon delivery. Tare weights [weight of the product after the weight of the packaging material has been subtracted] will be determined for each product and only net weights will be accepted rather than any printed weights on the packaging material. All other aspects of the specifications will be strictly enforced at all times and it is required that the purveyor cooperate to the fullest in preparing and shipping products for acceptance. All products shall be in sanitary condition, free from undesirable odors, blood clots, bruises, discolorations or other blemishes. In situations where such defects are found concealed within the product, the purveyor will be expected to make adjustments even after the product has been accepted.

Any product ordered in the fresh state shall be in excellent condition, possessing the quality, color, and other attributes typical of that grade and cut of meat. Items called for in the frozen state shall show no evidence of thawing and shall be packaged properly to prevent freezer burn. If for just cause, such as late delivery, lack of staff, or other such emergency conditions, a shipment of product is accepted without thorough inspection and found to be undesirable later, the purveyor will be expected to pick up the items on his next scheduled delivery date.

Should you have any questions concerning these policies and procedures please contact our food and beverage manager, Mr. Llenroc.

Sincerely,

Stephen A. Mutkoski

14.4 A POOR ATTITUDE TOWARD PURCHASING CREATES PROBLEMS

Historically, the purchasing department and in particular the job of the purchasing agent have been viewed as strictly clerical. Hence it is common to find the cook, the bartender, or even the hostess assuming the duties of purchasing agent. Their lack of expertise costs many food service operations thousands of dollars in excess food and beverage expense every year.

Profit Center

The purchasing department must be viewed as a profit center. If items are purchased economically, the money saved by proper buying procedures can be counted as profit. This is the attitude of most well-organized restaurant and hotel chains, and it is slowly filtering down to smaller, privately run restaurants. If the purchasing department can reduce its cost for a strip loin from $3.00 to $2.50 by competitive buying (to be covered later in this chapter), the 50¢ per pound saved should show up as additional profit. The menu price is already established, the labor cost remains the same, the overhead and preparation costs are exactly the same, and, therefore, the savings in purchasing price generates an increase in profit.

Understanding the Total Picture

The purchasing agent must be an individual who understands the goals of the food service operation. One must understand the market served: What level of quality does the clientele expect? The purchasing agent must be aware of the financial status of the food facility: Can the operation afford to purchase large quantities of product if the price is right, or is the cash flow so minimal that the product must be purchased on a week-to-week basis? The purchasing agent must know the storage capacities of both the operation and its various purveyors. Should it be necessary to hold merchandise for a substantial period of time, one must be aware of the costs for long-term storage. The purchasing agent should be someone with the ability to do research to determine product availability, market trends, and conditions. It is important for the purchasing agent to know whether a product such as South African lobster tails will be available in June or what the price range of beef is expected to be within the next three or four months. One must be aware of new products coming on the market that might increase the marketing base of the operation and relieve the demand for some of the menu items that cost the operation a lot to prepare.

Ethics

The purchasing agent must be very trustworthy and honest. Gifts, favors, and payoffs are commonplace in many areas of business. The individual chosen to perform the purchasing task must be an honest person with a high degree of professionalism. The purchasing agent's duty is to obtain the product most suited for the operation, at the time when it is needed and at the most reasonable cost. If a purchasing agent is getting a **kickback** of 1–5 percent of purchases, the purveyor will recoup the cost of the bribe from the customers supplied. This is most often accomplished by overcharging, short weighting, or shipping product of inferior quality. Management must safeguard against loss by establishing good receiving procedures (see the next chapter) and constant checks on the purchasing function to be sure that competitive buying is maintained.

14.5 MARKET INFORMATION – TRENDS AND PRICES

The purchasing agent must be aware of what's available, when, at what cost, and why. Among the best sources of information are sales representatives from the various purveyors. These people deal with the market daily and are often well informed on both short-term and long-term market conditions. They also provide new product information and samples for testing.

Many government and industry publications furnish very valuable information concerning market trends. Table 14.9 lists some of these publications.

To show how market publications can be beneficial to the purchasing agent, a few samples will be discussed here. First, consider the *Livestock Market News Weekly Summary Statistics*. Written on the inside cover of this periodical is a brief weekly summary of what is happening in the live cattle, hog, and sheep market. (See Figure 14.1).

The same publication provides graphs comparing the price structure for the current year with that for the previous year (see Figure 14.2). The price of live cattle affects the price the restaurant must pay for wholesale and portion cuts. Since there is something of a delay between the time the live cattle is purchased and the time it is processed, this graph is a good short-term indicator for price trends.

This publication also provides wholesale price quotations per railroad car lot for all of the major primal and wholesale cuts (see Table 14.10). Using these quotations, a purchasing agent is able to judge whether or not a purveyor is charging a fair price. (Most purveyors' gross profit is 15 percent on wholesale cuts.)

TABLE 14.9 SOURCES OF MARKET INFORMATION

Dairy

Dairy Market News
U.S. Department of Agriculture
Federal-State Market News Service
801 West Badger Road
Madison, Wisconsin 53713

Dairy Market News
U.S. Department of Agriculture
Agricultural Marketing Service
Dairy Division
970 Broad Street
Newark, New Jersey 07102

Livestock

Agricultural Prices
Crop Reporting Board
Statistical Reporting Service
U.S. Department of Agriculture
Washington, D.C. 20250

Market News
Livestock Division
Agricultural Marketing Service
U.S. Department of Agriculture
Washington, D.C. 20250

Cattle on Feed
Crop Reporting Board
Statistical Reporting Service
U.S. Department of Agriculture
Washington, D.C. 20250

Fish and Shellfish

Shrimp
U.S. Department of Commerce
National Oceanic and Atmospheric
   Administration
National Marine Fisheries Service
Washington, D.C. 20230

Industrial Fishery Products
Market Review and Outlook
U.S. Government Printing Office
Division of Public Documents
Washington, D.C. 20402

Shellfish
U.S. Government Printing Office
Division of Public Documents
Washington, D.C. 20402

Food Fish
U.S. Government Printing Office
Division of Public Documents
Washington, D.C. 20402

Fruits and Vegetables

Vegetable Situation
Economic Research Service
U.S. Department of Agriculture
Washington, D.C. 20250

Supply Guide
United Fresh Fruit and Vegetable
   Association
1019 19th Street, N.W.
Washington, D.C. 20036

Container Net Weights
United Fresh Fruit and Vegetable
   Association
1019 19th Street, N.W.
Washington, D.C. 20036

Western Growers Association
Vegetable and Melon Industry of
   California and Arizona
3091 Wilshire Blvd.
Los Angeles, California 90005

Monthly Supply Letter
United Fresh Fruit and Vegetable
   Association
1019 19th Street, N.W.
Washington, D.C. 20036

Fresh Approach Foodservice News-
   letter
United Fresh Fruit and Vegetable
   Association
Suite 1900
40 West 57th Street
New York, New York 10019

Fresh Facts for Foodservice
United Fresh Fruit and Vegetable
   Association
1019 19th Street, N.W.
Washington, D.C. 20036

Covers All Categories

Food Purchasing Guide for Group
   Feeding
Superintendent of Documents
U.S. Government Printing Office
Washington, D.C. 20402

Figure 14.1 Inside cover of Livestock Market News Weekly Summary and Statistics.

2   LIVESTOCK REVIEW

CATTLE: Prices for slaughter steers and heifers were rather uneven for the week. Most areas of the Nation showed upturns but the midwestern direct trade reported lower prices. Trade on slaughter steers and heifers was fairly active early in the week. However, it slowed at midweek as the wholesale beef carcass trade failed to show much support. Buyers continued to be very cautious and selective, especially toward yield grade. Yield grade continued to be a very important price-determining factor, with a substantial price spread between Yield Grade 3 and 4 carcasses. Demand for market-ready cattle tapered off for the Thanksgiving holiday and many sellers delayed marketing until after the holiday. Receipts of feeder cattle were sharply reduced for the holiday-shortened week and prices were sharply higher with most advance in the Southeast.

TRENDS: Slaughter steers and heifers steady to $2.00 higher, except the midwestern direct marketing areas were steady to $1.00 lower. Slaughter cows $1.00-2.00 higher. Slaughter bulls steady to $1.00 higher. Vealers at South St. Paul steady to $10.00 lower. Feeder steers and heifers steady to $2.00 higher with the Southeast steady to $3.00 higher.

PRICES: Slaughter steers: Choice 2-4 900-1300# east of the Rockies $63.00-67.25, west of the Rockies $66.00-72.00. Slaughter cows: Utility and Commercial 2-4 in the Midwest $43.00-47.00. Slaughter bulls: Yield Grade 1-2 1200-2100# $48.00-55.00. Vealers at South St. Paul: Choice and Prime 140-250# $80.00-90.00. Feeder steers: Medium Frame 1 300-500# $75.00-82.00, 600-700# $72.00-78.00 at the river markets.

HOGS: Prices for barrows and gilts were sharply higher for the week under review. Higher dressed pork trade, along with reduced receipts, forced live hog prices higher.

TRENDS: Barrows and gilts $1.00-2.00 higher. Sows steady to $1.00 higher. Feeder pigs per head mostly steady to $4.00 higher. Per hundredweight $2.00-16.00 higher.

PRICES: Barrows and gilts: US 1-2 200-240# at eastern terminals $48.00-49.00, western terminals $47.00-48.00. Directs $45.50-47.00. Sows US 1-3 300-600# $39.00-43.00 at terminals. At direct points $38.00-41.00. Feeder pigs per head: US 1-2 40-50# $32.00-49.00.

SHEEP: Prices for slaughter lambs were sharply higher for the week. Reduced marketings for the holiday-shortened week was the major price-supporting factor. San Angelo was out of the market due to the Thanksgiving holiday with no sales reported.

TRENDS: Slaughter lambs at Midwest market centers $2.00-4.00 higher. In direct trade areas steady to firm in Iowa-Southern Minnesota, and steady in Colorado. Slaughter ewes and feeder lambs in the Midwest steady to $1.00 higher.

PRICES: Slaughter lambs: Choice and Prime 95-120# wooled $48.00-60.75 at Midwest market centers. Shorn with No. 1 and 2 pelts $58.00-61.00, up to $63.00 at Sioux Falls. Slaughter ewes: Cull to Good in the Midwest $7.00-15.00. Feeder lambs: Choice and Fancy 60-90# in the Midwest $59.00-65.00.

                                               MEAT REVIEW

BEEF: Prices were uneven for steer and heifer carcasses, mostly lower except steer beef higher on the west coast. Steer and heifer beef carcasses opened the week in light demand and trading, especially to chainstores and processors. Most packers held firm on Monday but by Tuesday were forced to take lower prices in order to move carcasses for immediate shipments. Cow beef demand was good on limited holiday-shortened week supplies. Boneless beef advanced as retail demand was good. Heavy supplies of boxed beef found very limited demand as most retail outlets were concentrating on poultry products.

PORK: The fresh pork cut trade opened the week slow but turned active at midweek. Loins held mostly steady in heavy trading, with many shipments deferred until the weekend or early the next week. Some lightweight loins moved to higher price levels. Picnics and Boston butts were mostly steady in light trading. Hams continued to fall to lower price levels, mainly on weights over 17# due to burdensome inventory of frozen hams.

TRENDS: Steer beef $2.00 lower, except the west coast areas steady to $3.00 higher. Heifer beef uneven, mostly $2.00 lower. Cow beef $2.00-3.00 higher. Boneless beef $3.00-5.00 higher. Prime special fed veal $5.00 lower; Good and Choice boning veal $3.00-4.00 lower. Fresh pork loins 17# down steady to $4.00 higher, 17# up steady. Shoulders, picnics, Boston butts, and spareribs mostly steady, instances $1.00-2.00 higher on the west coast. Fresh hams 14-17# mostly $1.00-3.00 higher except $1.00-7.00 lower on the west coast; 17# and up $1.00-6.00 lower. Bellies mostly steady to $1.50 lower except west coast 50¢ higher. Lamb 55# down steady to $2.00 higher; 55# up $3.00-5.00 higher on the east coast.

PRICES: (Closing prices Central U.S. f.o.b. Omaha basis). Choice 3 steer beef 600-900# $99.00-100.00. Choice 3 heifer beef 500-700# $97.00. Cow beef Canner and Cutter 1-2 350# and up $91.00. Fresh pork loins 14-17# $88.50-92.00. Skinned hams 17-20# $83.00-83.25. Bellies 14-16# $61.50-62.50.

Source: Livestock Division, U.S. Department of Agriculture, Washington, D.C., vol. 48, no. 48, Dec. 2, 1980.

Figure 14.2 Price structure graph for two years.

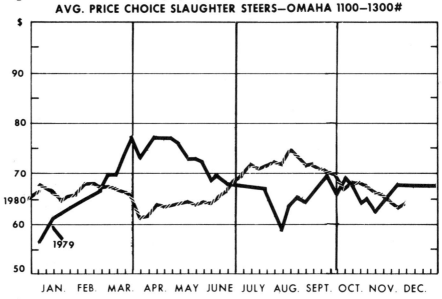

AVG. PRICE CHOICE SLAUGHTER STEERS—OMAHA 1100-1300#

TABLE 14.10 PRICE QUOTATIONS FOR MAJOR PRIMAL AND WHOLESALE CUTS

Weekly average of daily quotations in dollars per 100 lbs.

Carlot Basis		Dec. 1 1979	Nov. 29 1980	Dec. 1 1979	Nov. 29 1980	Dec. 1 1979	Nov. 29 1980
		East Coast CAF		Central U.S. FOB		Los Angeles CAF	
IMPS Item No.							
107 Rib, Oven Prepared	17-28#	--	--	195.00	--	--	--
109 Rib, Roast Ready	14-24#	--	--	233.75	203.38	236.88	212.67
112 Ribeye Roll	6-11#	--	--	--	--	460.83	427.50
112A Ribeye Roll, Lip on	7-12#	--	307.50	343.00	302.50	352.50	324.83
114 Shoulder Clod	13-24#	141.38	135.00	137.15	128.50	141.88	135.00
116A Chuck Roll	13-25#	--	149.00	146.50	142.42	--	160.00
120 Brisket(Bnls),Deckle off 6-14#		128.50	127.50	125.85	122.62	139.75	132.00
126 Armbone Chuck (Bnls)	60-90#	117.90	121.75	115.60	117.69	--	122.00
160 Round,Shank off,(P Bnls)45-80#		150.50	--	147.58	143.88	--	--
161 Round, Shank off(Bnls)	40-75#	161.00	161.00	157.50	156.00	--	--
167 Knuckle	8-13#	165.00	--	162.75	159.88	172.50	170.67
167A Knuckle (TRMD)	8-13#	--	--	182.00	--	--	--
168 Top(Inside) Round	14-23#	--	165.00	172.75	162.12	179.25	172.67
170 Bottom(Gooseneck)Round	18-29#	155.50	156.75	151.31	151.31	157.75	159.00
171 Bottom(Gooseneck)Round(Untrmd)18-29#		--	--	--	--	--	--
171A Bottom(Gooseneck)Round							
(Untrmd, heel out)	17-28#	--	--	--	--	--	--
173 Short Loin (TRMD)	17-30#	--	--	--	201.00	--	--
175 Strip Loin	11-24#	--	185.00	--	184.50	--	--
179 Strip Loin, Short Cut	8-14#	--	--	--	--	--	--
180 Strip Loin,Short Cut(Bnls)8-10#		--	--	--	--	--	287.50
	10-12#	--	--	298.50	--	302.50	270.33
	12# up	--	225.00	298.50	220.50	296.88	265.33
184 Top Sirloin Butt	7-10#	--	--	201.50	185.50	--	192.00
	10-12#	--	--	203.44	176.50	211.88	193.50
	12-14#	--	--	196.75	--	210.88	193.50
185 Bottom Sirloin Butt	5-7#	--	--	98.00	100.00	113.00	--
	7-9#	--	--	105.50	114.00	--	--
186 Bottom Sirloin Butt(TRMD)2-7#		--	--	--	--	--	--
189 Full Tenderloin	5-7#	--	300.00	--	285.50	307.50	288.00
185A Bottom Sirloin, Flap	1-3#	--	--	--	--	178.33	171.50
185B Bottom Sirloin, Ball Tip1-2#		--	--	--	--	--	--
	2# up	--	--	--	--	177.50	197.50
185C Bottom Sirloin, Triangle1½-5#		--	--	--	--	167.50	183.67

*Abbreviations are as follows: institutional meat purchasing specifications (IMPS); boneless (BNLS); trimmed (TRMD); free on board (FOB).

The following example illustrates the proper use of the chart shown in Table 14.10. A # 109 Roast-ready rib sells for $2.33/lb FOB on Dec. 1, 1980. Fifteen percent of that figure is 35¢, so the purveyor will sell the product for approximately $2.68.

CATTLE ON FEED   This monthly publication indicates the percentage of cattle placed on feed in any given month and generally indicates whether this is an increase or decrease from last year and last month. Since most food operations are dealing with Prime and Choice grades of beef, their supplies come from feedlot cattle rather than cattle that graze. Hence if there is a serious reduction in the number of cattle placed on feed for any given month, 2½–3 months later there will most definitely be a shortage of fed cattle for slaughter. The result may well be an increase in price for the top quality grades. Thus any indication of a drop in the number of feedlot cattle should signal the purchasing agent to purchase for future needs.

FISH AND SHELLFISH   Publications on fish and shellfish give the buyer some very valuable statistics. They generally list the amount of product in frozen storage and the amount of the estimated catch for the current season and compare these to the previous year's levels. These publications should be followed especially closely by purchasing agents dealing in costly commodities such as shrimp, lobster, and crab meat.

How Price Trends Can Be Predicted

It is possible to use the graphs provided in the *Livestock Market News* to show some market trends for different months of the year. These trends can also be developed from one's own purchasing records by keeping price quotes from week to week, month to month, and year to year.

Table 14.11 shows a form that can be used by the purchasing department to take price quotes from various purveyors.

Using statistics one gathers oneself (through price quotations) and from USDA and industry publications, one can see some price trends that generally hold true from year to year. Figures 14.3–14.6 show price trends on some of the high-priced beef items heavily utilized by the food service industry.

BEEF RIBS AND LOINS   Historically, beef loins and ribs (see Figures 14.3 and 14.4) fit into a very similar pattern, with the lowest prices being at the beginning of the year, from January through March. Then prices begin to rise until they reach a peak in the summer months, May through August. The price trend is downward through October and November, with generally a slight rise in price during the holiday season in December. This information would be valuable for operations doing peak business in the summer months. Caterers, who use a tremendous number of #109 ribs for wedding and graduation parties in May and June, could save substantially by purchasing in March and holding the product in freezer storage for two months. Steak house operations having a high de-

TABLE 14.11   FORM FOR OBTAINING PRICE QUOTATIONS FROM PURVEYORS

Our Spec #	NAMP #	Item	Grade	Wt. Range	Sam's Packing	Custom Meats	HRI Provisions	HEC Supplies
	Beef							
B 1034	# 103	Primal rib	Choice, yield 2	32–34				
B 1095	# 109	Roast-ready rib	Prime, yield 2	18–20				
B 1806	# 180	Strip loin	Choice, yield 2	10–21				
	Lamb							
L 2042	# 204	Hotel rack	Prime	6–8				
L 2323	# 232	Loin	Choice	8–10				
L 2334	# 233	Legs (pairs)	Choice	22–24				
	Veal							
V 13061	#1306	Rib chops	Nature fed					
V 3352	# 335	Leg, boneless	Nature fed					
V 3323	# 332	Loin	Nature fed					

mand for steaks such as the strip loin and tenderloin during the summer months might also use this information for purchasing at low price points throughout the year. Storage costs and interest costs must be considered (see Sec. 14.6, Methods of Buying).

BEEF ROUNDS   USDA Choice primal rounds vary little from month to month (see Figure 14.5). The purchasing agent need not follow price trends on such an item very closely. There is little to be gained in buying this item too far in advance of the need.

BEEF BRISKETS   When the loin and rib become extremely high in price due to heavy demand, the areas of the forequarter other than the rib are discounted and sold at significantly lower prices. To see this effect, imagine superimposing the fluctuating price structure of briskets graphed in Figure 14.6 on the price structures for beef ribs or beef ribs, shown in Figures 14.3 and 14.4, respectively. The cost curve for brisket would be the opposite of the cost curves for the other two cuts. The best time to buy briskets would be in the summer months and the most costly times would be in the winter months through March and April.

Figure 14.3   Price trends on 25/30 beef ribs, U.S. Choice.

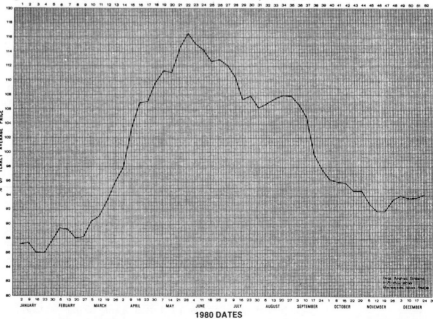

Source: Meat Price Relationships: A Guide to Estimating Seasonal Price Moves in Meat and Poultry Products. (Minneapolis: Price Analysis Systems, 1980). Reprinted by permission.

Figure 14.4   Price trends on trimmed 50/50 beef loins, U.S. Choice.

Source: Meat Price Relationships: A Guide to Estimating Seasonal Price Moves in Meat and Poultry Products. (Minneapolis: Price Analysis Systems, 1980). Reprinted by permission.

Figure 14.5 Price trends on steer rounds, U.S. Choice.

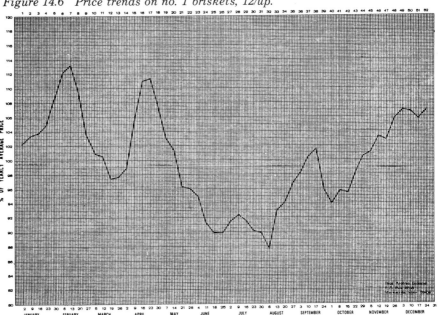

Source: Meat Price Relationships: A Guide to Estimating Seasonal Price Moves in Meat and Poultry Products. (Minneapolis: Price Analysis Systems, 1980). Reprinted by permission.

Figure 14.6 Price trends on no. 1 briskets, 12/up.

Source: Meat Price Relationships: A Guide to Estimating Seasonal Price Moves in Meat and Poultry Products. (Minneapolis: Price Analysis Systems, 1980). Reprinted by permission.

14.6 METHODS OF BUYING

It is possible to list many methods of buying. Five will be discussed in this section: **open market, competitive bid, contract purchasing, bulk buying,** and **futures.**

Open Market

The buying method that does not restrict the purchasing agent to any particular source of supply is called open market. The technique may be as rudimentary as going to the marketplace and selecting the product from various vendors. The term generally refers to a system of buying in which the purchasing agent either calls or receives visits two or three times a week or even daily from salesmen representing several purveyors in order to obtain price quotes on which to base the decision of which supplier to use. Previous experience with various suppliers also influences the choice. The majority of small restaurant operations use this system for purchasing most meat, fish, and poultry items.

Competitive Bid

The open market system can be taken one step further to guarantee the lowest possible cost for comparable product quality and service. Each time a supply of products is needed a purchasing agent using a competitive bidding system contacts two or three suppliers (ones the agent has already surveyed and approved for business) to quote prices on the items needed. The order is awarded to the firm with the best competitive bid. This method of buying allows the purchasing agent to consider quality and consistency of product and service without contractual agreement with any one source of supply. Competitive bidding is more time consuming than open market purchasing and is generally used only on higher cost items—meat but not bread.

Contract Purchasing

Some firms contract for a given quantity of meat products over a period of time (usually three to six months). The contract states a particular price for X quantity of product for the duration of the contract. This buying method thus assures the food service operation of a supply of product at a stated price for present and future needs. It is also advantageous to the supplier, who is guaranteed a sale of X quantity of product and does not have to worry about the possibility of losing the sale every week as may be the case in competitive bidding or open market purchasing. The system is used quite effectively by many large chain operations. Its weakness is that the prices of products remain fixed by contract whereas the market price may drop significantly after the contract is signed. The food service operation may thus have to pay a higher price for the commodity than if it were bought on an open market or competitive bidding basis. Some contracts are written with an adjustment clause that alters the price on a monthly basis according to the market fluctuation. This reduces the risk to both the food service operator and the purveyor (in case prices rise).

Bulk Buying

Purchasing agents with a good background in market trends and conditions may be able to judge low market points when the food operation could save money by buying a large quantity of merchandise and storing it for future use. Storage can be accomplished on the premises if facilities are available, or a storage fee may be paid to a purveyor or freezer plant to hold the merchandise until it is needed. A number of factors must be considered when deciding to buy a large quantity of merchandise for future use. First, the ability of the purchasing agent to judge the market trends must be reviewed. The average market price, the predicted low market price, the predicted high market price, and the special buy price must be determined. The cost of borrowing or tieing up capital funds must be considered and the storage costs either for electricity usage on the premises or for rental space must be figured into the cost analysis. The following example showing how bulk buying can benefit a food operation is provided for your information.

BULK BUYING PROBLEM—POTENTIAL SAVINGS COMPARED TO DAY-TO-DAY BUYING

Part A. Storage Cost

Given
1. A restaurant uses 24,000 lb/year (or 2,000 lb/month) of NAMP #180 strip loin.
2. Purchase price for #180 strip loin
   a. Predicted low price   = $2.85/lb
   b. Predicted high price   = $3.85/lb
   c. Yearly average price   = $3.35/lb
   d. Special price for bulk purchasing = $2.65/lb

Step 1. Multiply yearly usage by special purchase price.

$$24,000 \text{ lb/year}$$
$$\times \ \$2.65$$
$$\overline{\$63,600}$$

Step 2. Multiply yearly usage by average purchase price.

$$24,000 \text{ lb/year}$$
$$\times \ \$3.35$$
$$\overline{\$80,400}$$

Step 3. Subtract yearly special purchase price from average yearly price to determine the potential savings per year.

$$\begin{array}{r} \$80,400 \\ -\$63,600 \\ \hline \$16,800 \end{array}$$

If the restaurant uses 2,000 lb/month, then it will cost $5300/month (plus interest and storage cost).

Part B. Interest Cost (either for money borrowed to buy the product or interest lost on money withdrawn from the purchaser's savings to buy the product).

Given
1. Interest cost = 15% annually
2. Storage cost = $0.04/lb/month
(covers storage and security cost.)

Note: Estimate that the product is used in almost equal amounts over the year, so the amount stored, and hence the storage cost, is reduced by $\frac{1}{12}$ each month.

Step 1. Determine the storage cost by multiplying the number of pounds in storage each month by the cost of storage per pound per month. For example,

$$24,000 \text{ lb} - 2,000 \text{ lb/month} = 22,000 \text{ lb the first month}$$
$$\times \$0.04 = \$880.$$

The annual storage cost is computed by subtracting 2,000 lb from the preceding month's amount of product stored and multiplying the new amount by the storage cost per pound, then adding all the monthly totals for twelve months (see Table 14.12).

TABLE 14.12 COMPUTING ANNUAL STORAGE COSTS

Month	Amount Stored		Storage Cost lb/month		Monthly Storage Total
1	22,000	×	$0.04	=	$880
2	20,000	×	0.04	=	800
3	18,000	×	0.04	=	720
4	16,000	×	0.04	=	640
5	14,000	×	0.04	=	560
6	12,000	×	0.04	=	480
7	10,000	×	0.04	=	400
8	8,000	×	0.04	=	320
9	6,000	×	0.04	=	240
10	4,000	×	0.04	=	160
11	2,000	×	0.04	=	80
12	...				0
				Annual storage cost	$5,280

TABLE 14.13 COMPUTING ANNUAL INTEREST COSTS FOR BULK BUYING

Month	Purchase Cost Outstanding		Interest Rate/month		Monthly Interest Cost
1	$63,600	×	15% × $\frac{1}{12}$	=	$795
2	58,300	×	,,	=	729
3	53,000	×	,,	=	663
4	47,700	×	,,	=	596
5	42,400	×	,,	=	530
6	37,100	×	,,	=	464
7	31,800	×	,,	=	398
8	26,500	×	,,	=	331
9	21,200	×	,,	=	265
10	15,900	×	,,	=	199
11	10,600	×	,,	=	132
12	5,300	×	,,	=	66
			Annual interest cost		$5,168

Step 2. Determine the interest costs by multiplying the purchase price (less the $ amount used per month) times interest rate per month (15% × $\frac{1}{12}$). For example,

$63,600 (total purchase price) – $5,300 (monthly usage)
= $58,300 base to compute interest cost for the first month.

$58,300 × 15% × $\frac{1}{12}$ = $729.

To compute the annual interest cost, compute the monthly interest and add the amounts for 12 months, as shown in Table 14.13.

Part C. Savings

Given
1. Potential savings/year = $16,800
2. Storage + Interest costs = $10,468

Step 1. Determine actual savings by subtracting the storage and interest costs from the potential savings.

$16,800
–$10,468
———————
$ 6,332 savings

Note: This savings drops directly to the bottom line (since your menu and other costs remain the same) and is *profit.*

Step 2. Determine savings per pound by dividing the savings by total pounds.

$$\frac{\$6,332}{24,000} = 26.4¢/lb$$

Step 3. Determine actual cost per pound by adding the special purchase price plus the storage and interest costs and then divide the sum by the total pounds used per year.

$$\begin{array}{l} \$63,600 \\ +\$10,468 \\ \hline \$74,068 \end{array} \qquad \dfrac{\$74,068}{24,000 \text{ lb}} = \$3.09 \text{ (Actual cost/lb)}$$

Therefore, the market price must average less than \$3.09/lb in order for this purchase to be unprofitable. **Consider:** If you sell a strip steak for \$9.00 à la carte and you hit the national average of 5% net profit on food sales, you will make \$0.45 on each steak. Therefore, you must sell roughly 14,070 steaks before your profit reaches the same amount (\$6,332) as the savings made on bulk purchase of the amount of meat purchased in this problem.

Commodities or Futures Market Buying

With the advent of highly professional purchasing departments in both individual and chain restaurants, the futures market is coming into more use as an alternative buying method. In order to understand this method, it is necessary to have some general information concerning the commodities market and the people who make it work.

The futures market is similar in some ways to the stock market. Buying and selling occurs at the commodities exchange as it does at the stock exchange, but the commodities bought and sold are not available at the time of transaction, hence the term futures market. In effect, the individual who is selling the commodity promises to deliver it at the agreed price at a specific time in the future (when the crop is harvested or the animals slaughtered). The individual who is buying agrees to take delivery of the commodity at the future time and pay then the price agreed to at the time of the transaction. Table 14.14 reproduces the futures prices listed in the *Wall Street Journal* for several commodities.

Who Makes the Futures Market Work Traders on the futures or commodities exchange represent a variety of interests: food growers, food processors, and food purchasers. The first category, producers of crops and raw materials, trade in futures to guarantee a certain price for their product prior to harvest. A producer of soybeans might offer for sale a futures contract for 100,000 bushels of soybeans on January 1 with delivery on June 1 at a price per bushel of \$5.00. Although the product is not harvested yet, the producer has estimated all costs and is satisfied that there will be profit at the January price when delivery is made in June.

TABLE 14.14 EXAMPLES OF FUTURE PRICES AS LISTED IN THE *WALL STREET JOURNAL*, FRIDAY, JANUARY 9, 1981

	Open	High	Low	Settle	Change	Lifetime High	Lifetime Low	Open Interest
–GRAINS AND OILSEEDS–								
CORN (CBT)–5,000 bu.; cents per bu.								
Mr81	377	377¾	374½	374½	– 4	410¼	299½	111,737
May	385	385	381¼	381½	– 4½	418	307	57,427
July	386½	387	383¾	384	– 4	419¼	312	49,087
Sept	378	378¾	376½	376¾	– 2¾	407	331½	19,223
Dec	370	371¼	368¾	370¼	– 1¼	389	332½	20,742
Mr82	381	382	380	380¾	– 1¼	383½	369	182
Est vol 33,642; vol Wed 42,568; open int 258,398, +2,406.								
OATS (CBT)–5,000 bu.; cents per bu.								
Mr81	223	224¾	221½	222¼	– 1¾	242¼	168½	3,070
May	228	229¾	226¾	227½	– 1½	244½	179½	1,718
July	226	226¾	225	225	– 1¼	241	197	748
Sept	222	222½	221	221	– 1	237	205	206
Dec	224	224	224	224	– 1¼	227	218	103
Est vol 777; vol Wed 966; open int 5,845, +26.								
SOYBEANS (CBT)–5,000 bu.; cents per bu.								
Jn81	805	809	796	798	– 8	958	649	6,945
Mar	830	834	820	821¾	–12¼	988	667	61,998
May	854	859	846	847½	–10¾	1006	680	44,951
July	868	876	864	865	–10¾	1024	689	35,699
Aug	865	871	858	858½	–13½	1003	769	6,607
Sept	850	852	844	844½	– 8	933	755½	2,686
Nov	841½	843½	837	837¼	– 8¼	900	749	11,914
Jn82	858	863	856	856	– 7	916	794	779
Est vol 57,918; vol Wed 84,458; open int 171,579, –7,378.								
SOYBEAN MEAL (CBT)–100 tons; $ per ton.								
Jan81	234.00	234.50	230.80	231.50	– 3.50	294.00	181.50	2,919
Mar	243.50	243.50	240.00	240.50	– 3.70	300.50	185.80	23,021
May	250.00	250.50	246.70	247.30	– 3.60	300.50	190.30	14,671
July	255.00	255.20	251.50	252.50	– 3.80	301.50	194.00	8,698
Aug	253.50	254.00	251.50	252.30	– 2.90	295.00	213.50	2,175
Sept	248.00	249.50	248.00	249.00	– .50	278.50	213.50	1,621
Oct	242.00	243.50	242.00	242.70	....	262.00	216.30	2,485
Dec	242.00	244.00	242.00	243.00	....	259.00	218.00	2,436
Jan82	277.50	279.50	275.00	247.00	– .50	259.00	224.00	65
May	282.00	282.00	278.00	278.00	....	282.00	278.00	0
Est vol 9,282; vol Wed 17,095; open int 58,091, –791.								
SOYBEAN OIL (CBT)–60,000 lbs.; cents per lb.								
Jan81	25.00	25.10	24.80	24.90	– .25	29.65	21.55	4,486
Mar	25.60	25.85	25.50	25.57	– .26	30.60	21.90	28,574
May	26.50	26.65	26.30	26.33	– .34	32.00	22.72	12,138
July	27.15	27.30	27.05	27.07	– .31	32.00	22.72	10,742
Aug	27.25	27.45	27.15	27.15	– .30	31.60	23.25	2,487
Sept	27.45	27.55	27.30	27.35	– .25	31.20	25.97	1,688
Oct	27.55	27.70	27.35	27.45	– .20	30.05	25.62	2,516
Dec	27.75	27.85	27.50	27.58	– .15	29.73	25.40	2,458
Jan82	27.75	27.95	27.50	27.52	– .10	29.65	25.57	515
Mar	28.20	28.20	27.80	27.80	....	28.20	27.20	0
Est vol 8,410; vol Wed 13,759; open int 65,604, –1,261.								
WHEAT (CBT)–5,000 bu.; cents per bu.								
Mar	503	503	496	497¼	– 9	566½	433	29,944
May	509	510	505	505¾	– 7½	570½	440½	15,357
July	495	495½	489	489	– 9¼	552	447	17,530
Sept	508	508	500½	500½	– 9½	558½	461	3,883
Dec	526	526	515½	515½	–12½	569	485	2,070
Est vol 13,703; vol Wed 18,189; open int 68,784, +838.								
WHEAT (KC)–5,000 bu.; cents per bu.								
Mar	480	480	475	475½	– 5½	540	420	17,667
May	487	487	481	483¼	– 4¾	548½	429	4,444
July	487	487	481	481	– 6½	544	430	4,483
Sept	496	497	492	492	– 5	550	449½	196
Dec	510	510	510	510	– 1	511	421¼	0
Est vol 4,212; vol Wed 4,597; open int 26,786, –782.								
WHEAT (MPLS)–5,000 bu.; cents per bu.								
Mar	474	474½	468	468¾	– 6½	533¼	420	3,498
May	486	486	480½	481	– 6¾	543½	450	1,060
July	486½	490	482	482	– 8	544½	461½	277
Sept	495	496	492	492	– 8	545	478	33
Est vol 560; vol Wed 1,052; open int 4,869, –54.								
BARLEY (WPG)–20 metric tons; Can$ per ton								
Mr81	158.50	158.50	156.70	157.50	– 2.00	165.10	110.20	3,409
May	159.00	159.50	157.60	158.40	– 1.60	166.50	121.40	4,597
July	161.00	161.00	159.70	160.30	– 1.70	165.50	137.00	4,742
Oct	162.00	162.30	160.70	160.80	– 2.40	166.40	144.50	1,358
Dec	....	....	....	160.70	– 2.30	165.20	165.00	5
Est vol 2,100; vol Wed 1,466; open int 14,111, +131.								
FLAXSEED (WPG)–20 metric tons; Can$ per ton								
My81	430.50	430.50	421.80	421.80	–10.00	470.50	355.10	7,232
July	428.00	429.50	419.50	419.70	– 9.80	463.00	388.50	3,944
Oct	424.30	426.00	416.50	416.60	– 9.90	455.00	382.00	697
Est vol 1,420; vol Wed 779; open int 11,873, +111.								
RAPESEED (WPG)–20 metric tons; Can$ per ton								
Jan81	340.00	340.00	337.00	337.00	– 3.50	381.00	302.50	850
Mar	350.00	351.50	346.50	346.50	– 3.60	389.00	313.60	11,852
June	364.80	368.00	363.50	363.50	– 3.70	391.00	325.50	4,212
Sept	381.00	381.00	375.50	375.50	– 4.50	393.50	351.00	1,225
Nov	....	....	....	380.00	– 3.90	392.00	378.50	293
Est vol 850; vol Wed 1,170; open int 18,432, –106.								
RYE (WPG)–20 metric tons; Can. $ per ton								
May	240.50	241.40	239.00	239.00	– 2.70	250.50	136.00	1,719
July	228.00	228.10	227.00	227.00	– .70	237.80	185.00	1,076
Oct	....	....	....	201.80	– 2.10	213.00	188.00	493
Est vol 562; vol Wed 433; open int 3,288, –36.								
–LIVESTOCK & MEAT–								
CATTLE–FEEDER (CME)–42,000 lbs.; cents per lb.								
Jan81	72.85	72.85	71.75	71.85	– 1.00	82.85	69.50	238
Mar	75.65	75.95	74.82	74.85	– 1.25	84.25	72.00	3,839
Apr	76.45	76.75	75.70	75.75	– 1.17	84.30	70.00	2,362
May	76.45	76.77	75.75	75.82	– .92	84.05	72.85	1,789
Aug	77.30	77.30	76.40	76.45	– 1.22	83.00	73.70	407
Sept	76.70	76.70	75.80	75.80	– 1.00	82.40	73.30	263
Oct	76.30	76.50	75.70	75.70	– .80	80.10	73.50	147
Nov	77.00	77.00	76.20	76.20	no comp	77.00	76.20	0
Est vol 1,650; vol Wed 2,128; open int 9,045, unch.								
CATTLE–LIVE (CME)–40,000 lbs.; cents per lb.								
Jan81	64.00	64.90	64.00	64.50	+ .32	73.13	61.00	42
Feb	67.10	67.22	66.45	66.55	– .77	74.50	61.40	18,293
Apr	69.70	70.20	69.55	69.67	– .42	75.90	62.60	15,591
June	71.75	72.15	71.42	71.47	– .55	76.60	69.40	8,652
Aug	71.80	72.30	71.60	71.70	– .45	76.25	69.75	2,906
Oct	70.80	71.10	70.50	70.52	– .42	75.20	68.40	1,849
Dec	71.85	72.00	71.50	71.50	– .35	75.85	69.35	152
Est vol 18,256; vol Wed 18,963; open int 47,485, +498.								
HOGS (CME)–30,000 lbs.; cents per lb.								
Feb81	47.95	48.52	47.60	47.85	– .60	58.25	40.15	10,809
Apr	48.85	49.70	48.52	48.75	– .65	59.05	39.40	8,822
June	52.75	53.42	52.25	52.27	– .92	63.25	42.40	4,345
July	54.02	54.50	53.55	53.55	– 1.05	64.05	43.00	2,491
Aug	53.50	54.00	53.00	53.00	– .90	62.70	47.30	670
Oct	52.75	53.00	52.02	52.05	– 1.00	61.35	43.00	243
Dec	54.85	55.00	54.42	54.42	– 1.07	61.75	53.00	111
Feb82	....	....	....	57.25	....	61.85	55.60	16
Apr	....	....	....	57.25	....	61.50	60.90	1
Est vol 9,256; vol Wed 10,516; open int 27,508, +953.								
PORK BELLIES (CME)–38,000 lbs.; cents per lb.								
Feb81	55.10	56.00	54.05	54.12	– 1.67	74.50	36.60	8,512
Mar	56.10	56.95	55.00	55.07	– 1.70	75.00	38.75	5,578
May	58.50	59.05	57.20	57.25	– 1.72	77.05	41.42	3,693
July	60.35	61.20	59.30	59.35	– 1.50	79.05	42.50	3,003
Aug	60.30	61.00	59.05	59.15	– 1.42	78.40	42.00	1,113
Est vol 8,755; vol Wed 9,279; open int 21,899, –198.								
–FOOD & FIBER–								
COCOA (CSCE)–10 metric tons; $ per ton.								
Mr81	1,995	1,997	1,966	1,979	– 31	3,395	1,905	6,316
May	2,030	2,030	2,000	2,011	– 31	2,690	1,961	3,128
July	2,070	2,070	2,048	2,055	– 30	2,750	2,005	1,142
Sept	2,125	2,125	2,095	2,100	– 35	2,745	2,055	681
Dec	2,180	2,180	2,152	2,105	– 85	2,595	2,105	540
Mr82	2,225	2,209	2,215	2,215	– 30	2,280	2,170	35
Est vol 2,000; vol Wed 1,609; open int 11,842, –240.								
COFFEE (CSCE)–37,500 lbs.; cents per lb.								
Mr81	129.00	129.50	127.75	127.95	– 1.90	199.00	110.75	3,766
May	130.50	131.25	129.75	130.48	– .92	196.50	114.75	2,496
July	131.76	132.25	131.00	131.75	– .75	196.75	117.50	1,256
Sept	132.70	133.85	132.50	133.44	+ .19	198.00	119.00	862
Dec	132.75	132.75	132.75	133.25	– .35	165.50	120.00	309
Mr82	133.00	133.00	132.00	132.00	– .50	133.00	122.00	57
May	....	....	....	135.25	+ .75	....	....	2
Est vol 1,809; vol Wed 2,136; open int 8,748, –6.								
COTTON (CTN)–50,000 lbs.; cents per lb.								
Mar	93.50	94.35	93.30	93.88	+ .60	97.00	70.00	18,733
May	94.30	95.30	94.15	94.85	+ .60	97.67	72.80	8,208
July	94.03	94.80	93.85	94.65	+ .75	97.05	74.25	5,878
Oct	88.70	89.00	88.70	89.00	+ .20	89.95	75.50	1,131
Dec	85.70	85.70	85.40	85.51	– .08	86.10	75.80	4,138
Mr82	....	....	....	86.50	+ .25	86.50	81.47	84
May	....	....	....	87.50	– .25	87.00	84.70	4
Est vol 8,200; vol Wed 9,855; open int 38,176, +658.								
ORANGE JUICE (CTN)–15,000 lbs.; cents per lb.								
Jan	79.40	79.40	78.05	78.60	– .80	107.75	77.00	429
Mar	83.75	83.75	82.35	82.90	– .65	107.80	81.60	5,981
May	85.90	85.95	84.90	85.45	– .50	107.90	84.00	2,102
July	88.65	88.65	87.40	87.90	– .75	107.50	86.30	1,200
Sept	89.95	90.15	89.95	90.15	– .75	98.00	89.95	255
Nov	....	....	....	91.30	+ .40	99.50	92.70	16
Jan82	92.50	92.50	92.15	92.60	– .40	105.10	91.50	339
Mar	93.20	93.20	93.20	93.60	– .40	104.00	93.20	47
Est vol 1,000; vol Wed 1,189; open int 10,379, +40.								
POTATOES (NYM)–50,000 lbs.; cents per lb.								
Mar81	14.35	14.35	14.05	14.08	– .16	14.55	7.60	2,507
Apr	17.10	17.10	16.80	16.89	– .12	17.10	8.25	8,940
Nov	8.88	8.88	8.70	8.75	– .08	9.07	8.03	296
Est vol 2,108; vol Wed 2,502; open int 11,743, –33.								
SUGAR–WORLD (CSCE)–112,000 lbs.; cents per lb.								
Mar81	32.25	32.50	31.70	31.80	– 1.46	45.75	13.80	20,504
May	32.00	32.45	31.70	31.75	– 1.32	45.30	15.55	14,667
July	31.60	32.00	31.47	31.47	– 1.00	44.34	21.85	11,127
Sept	29.95	30.30	29.75	29.75	– 1.00	41.50	26.68	5,314
Oct	29.50	30.00	29.10	29.19	– .91	40.90	26.31	9,491
Jan82	26.75	26.75	26.75	26.75	– 1.00	37.80	24.00	278
Mar	26.25	26.60	26.00	26.06	– .84	36.60	24.13	3,318
May	....	....	....	25.91	– .84	26.50	26.40	114
Est vol 9,088; vol Wed 9,715; open int 64,813, +511.								

The second category of traders on the futures market are end users of a particular commodity who want to guarantee a supply of that commodity at a given price. A manufacturer of meat analogues, for instance, might contract with the producer of soybeans to have a guaranteed supply of the product at a fixed price. The manufacturer would purchase the futures contract stating that delivery of 100,000 bushels of soybeans will be taken in June at $5.00 per bushel. Advantages accrue to both sides. The producer does not have to worry about the price falling in April or May because he has a guaranteed price per bushel for June delivery. The end user is guaranteed a supply but does not have to tie up $500,000 plus storage, insurance, and other costs for a six-month period of time, as he would have to do if he bought and stored the product for future use. He is guaranteed a certain price, cost per bushel, and is tieing up only a small amount of money in the form of the "margin" required to negotiate the contract. The size of the percentage or margin depends on the commodity; for example, a margin of only $1,000 would be required on a contract for nearly 40,000 pounds of pork bellies, whereas the margin on the same amount of beef would be $1,200.

The third type of user of the commodities market is trying to hedge against future price increases. Purchasing agents for the food service industry belong to this category. A large school feeding operation, using hundreds of pounds of beef over a year, might elect to hedge against price increases by purchasing a number of futures contracts for live cattle. If in January the price of hamburger is $1.00/lb and the school's purchasing agent knows that it is predicted that prices will increase over the next six months, the agent can avoid this price increase by purchasing in advance a certain number of futures contracts for live cattle spread out over a period of six months to one year.

Since ground beef is processed from beef cattle, it is reasonable to expect that an increase in live cattle prices would cause an increase in ground beef prices. The following example shows how effective futures trading can be as a hedge against adverse market turns.

Given:
- Estimated usage of 200,000 lb ground beef for the year
- Current cash price of $1.00/lb
- Five futures contracts for live cattle (totaling 40,000 lb of beef each) purchased for delivery in each of five months at prices shown.

February	April	September	November	December
51¢	53¢	55¢	57¢	59¢

As the ground beef is actually needed, it is purchased on the cash market and a futures contract is sold to counter market price fluctuations. If live cattle prices increase as February approaches, the cost of ground beef goes up. The increased cost of purchasing the actual commodity is offset by the sale of the now higher valued futures contract.

TABLE 14.15 TRADING IN FUTURES OF BEEF

| | Futures Market Prices per Pound | | | | |
	February	April	September	November	December
Futures contract price when purchased in January	$0.51	$0.53	$0.55	$0.57	$0.59
Futures contract price when sold	0.54	0.51	0.57	0.55	0.59
Gain or loss	+0.3	−0.02	+0.02	−0.02	0
	Cash Market Prices per Pound				
Purchase cost of ground beef	$1.03	$0.98	$1.02	$0.98	$1.00
Gain or loss from January base of $1.00	−0.03	+0.02	−0.02	+0.02	0

If live cattle prices go down, a loss is taken on the futures contract, but it is offset by lower costs for the ground beef on the cash market. Hedging is an attempt to stabilize cost for future periods of time while minimizing risk. What should happen to the five contracts purchased and the cost for ground beef theoretically during this time period is reflected in Table 14.15.

The net effect of futures trading in the foregoing example is that gains offset the losses to yield a stable cost of $1.00/lb throughout the year. To buy 200,000 lb of ground beef in bulk would require a significant investment of funds plus interest and storage costs whereas purchasing the futures contract for hedging requires only a small investment on the margin. Since the futures contract is being sold only when an actual cash commodity is purchased, there is very little risk involved in this method of buying.

The final category of traders in the commodities market is that of speculator. The speculator seeks to make a profit by estimating which way the price of a commodity is going to move. If the person believes that soybean prices will go from $5 to $6 per bushel within three to six months, he or she would purchase a certain number of soybean contracts and hold them for future sale. If the market moves as predicted, the speculator stands to profit substantially from the increased value of his contracts. The risks involved are generally limited to the amount of margin put up by the speculator to open the contract.

Since the commodities market is based on future transactions, the speculator may also start trading by selling a contract if he or she predicts the prices will decline in future months. At that point in time, the contract would be purchased back and a profit realized. Speculators are an important part of the commodities market and in fact are the ones that create a great deal of liquidity for the market and give the hedger, in the preceding example, the opportunnity to buy and sell at will.

14.7 SUMMARY

Purchasing is the first link in the flow of food products through the food service system. It is a critical area because it limits the performance of the subsystems (receiving, storage, preparation) that follow it. If the purchasing department does not function properly in defining the need through its purchase specification, in communicating the need, in selecting adequate sources of supply, and in using the most effective buying methods, the financial outcome of the food operation will suffer. All the efforts put in at the receiving dock and in the storage and preparation areas cannot totally correct a poor purchasing job. Quality and efficiency must start at the beginning of the chain.

CASE STUDY You have been hired by the Blanc Hotel Corporation to set standards for their national purchasing network. Blanc maintains an image of quality in its hotels throughout the country. The menus differ from one hotel to the next, but there are several standard items, including the following:

- Prime Ribs of Beef
- Chateaubriand
- Rack of Lamb
- Veal Cordon Bleu
- Roast Capon
- Shrimp Scampi

You are asked to prepare a cover letter and a complete set of purchasing specifications for the menu items listed. Since this is a national chain, guidelines will have to be established for selection of local suppliers that can meet the requirements of Blanc Hotel Corporation. Discuss your policies and recommendations for the evaluation of these suppliers.

---

15 RECEIVING

Receiving is a key control point in a food and beverage operation. When functioning properly, the receiving department provides checks and balances that make the purchasing department accountable for its actions (or lack of action). Merchandise is money, and it is as important to be assured of every penny at the receiving dock as it is to balance the cash bank in the registers. Carelessness in either place will produce the same result—financial loss.

Functions of the Receiving Department

1. Ensure product quality, quantity, and condition
2. Aid the purchasing department in evaluation of suppliers' performance
3. Check the performance of the purchasing agent
4. Serve as an accounting checkpoint by providing a verified invoice to compare to billing statements
5. Maintain product condition by minimizing delays between receipt and production or storage
6. Inform purchasing and production of shortages resulting from incomplete deliveries or returns

15.1 THE KNOWLEDGABLE RECEIVING CLERK

Weak Link in the Chain?

Purveyors of meat, fish, and poultry interviewed throughout the United States claim that less than 10 percent of their food service customers do an adequate job of receiving. Some go through the motions, but many make no attempt at all to check the product upon delivery.

Managements often go to great lengths to develop a food service operation—conducting extensive market studies, hiring professionals to plan the menu, preparing purchasing policy and procedure manuals, writing purchase specifications—then proceed to let the dishwasher or potman accept delivery of the most expensive commodity, meat. Often such employees lack the fundamental knowledge and technical expertise to do an effective job. The person who serves as receiving clerk must have the knowledge of product characteristics for identification and quality evaluations. To be effective the receiving clerk must be able to distinguish between a tenderloin and a chuck roll or between USDA Commercial grade and USDA Choice.

Good Receiving Practices Start with Trained Personnel

Larger operations have the luxury of developing a receiving department with full-time receiving personnel. Most smaller operations rely on an individual who has other job functions as well as receiving, serving also as storeroom clerk, assistant manager, or cook, for example. Whether full-time or part-time the receiving clerk must be trained in the basic skills of receiving. This training combined with proper equipment, procedures, and information can make a dishwasher or potwasher into an effective receiving clerk.

SKILLS REQUIRED BY THE RECEIVING CLERK

Knowledge of Product Characteristics For identification of product received, the basic characteristics of all the major items purchased should be committed to memory by the receiving clerk. This book's chapters on quality and yield and descriptions and photos of each type of meat, fish, and poultry and other products should be helpful. A tour of a meat packing plant can also aid in developing identification skills.

Knowledge of Product Quality The grading terminology must be known for each commodity, but, more important, the visible characteristics used to judge grade should be understood by the receiving clerk. This knowledge gives the receiving clerk the ability not only to read a grade stamp and know what it signifies, but also to judge whether the stamp accurately reflects the quality of the product.

Knowledge of Product Weight Ranges and Trim Factors Much information is provided in the purchasing specifications, but the receiving clerk should familiarize himself or herself with standard weight ranges and trim factors for given products. Gross discrepancies on weight ranges and even minor discrepancies on trim can be very costly in the long run.

Knowledge of Acceptable Product Condition The receiving clerk should be cognizant of the conditions which acceptable products should possess: ideal color and temperature; freedom from slime, bruises, mold growth; and so on.

Knowledge of Packaging Material The receiving clerk should be able to distinguish between different types of packaging material and evaluate the condition of the material and the packaged product.

15.2 NECESSITIES AT THE RECEIVING DOCK

Facilities and Equipment

An area of the food service operation should be designated as the place to receive and inspect incoming merchandise. The space allotted to the receiving department obviously will vary with the size and scope of each operation. No matter how small, space must be allocated for receiving. All of the equipment and information required to perform the receiving function should be in this area.

Accurate scales are needed to weigh the products being purchased. Most operations should have at least two scales: one a floor or table model with a capacity of about 125 pounds for checking wholesale cuts and net weights and one or more portion scales for checking accuracy of individual portions. If the operation is purchasing carcasses, forequarters, or hindquarters, a rail scale is required. All scales should be inspected periodically for accuracy.

Other equipment needed includes table(s) for inspection, adequate lighting, moving equipment (hand trucks or carts), thermometers for checking internal temperature, and possibly a sink for draining moisture or ice off some products prior to weighing, a ruler for checking trim factors on certain meat items, and a fat analyzer for testing fat content in ground beef. Larger operations should provide a desk for the receiving personnel to handle paperwork and to file reference material.

Information to Be Provided to the Receiving Department

The receiving clerk should be given a copy of the purchasing specifications and possibly a copy of the purchase order (depending on the type of receiving system used). Also on hand should be purveyors' delivery schedules, and reference material such as the *Meat Buyers Guide*, this textbook, and charts to aid in product identification.

15.3 RECEIVING PROCEDURES

1. Request an invoice from the delivery man and compare it to the purchase order. Check to be sure quantities and prices are in agreement on both.
2. Determine the net weight for each individual item. Initially, this requires recording the gross weight of the boxed products, then unboxing and unwrapping the products to measure net weights accurately. The net weight is then subtracted from the gross to determine the "tare" weights for cartons and packaging material. Once the tare weights are known, this amount can just be subtracted from the gross weight on the scales to check net weights of items re-

ceived subsequently. For example, a case of strip loins containing six strips comes in a cardboard container that weighs 2½ lb and plastic wrap for each strip loin weighs 4 oz (6 × 4 = 24 oz). When the box is weighed, the gross weight is 70 lb on the scale, but the net weight determined by subtracting the tare weights determined previously is 66 lb(70 − (2½ + 1½)). One pays for the service of packaging in the price per pound quoted. Be sure your operation is not overpaying by mistaking the gross, packaged weight for the net weight it has ordered.

3. Check weight tolerances and ranges against the purchasing specification. Wholesale cuts are generally allowed a 2-lb tolerance and portion cuts ¼- to 1-oz tolerance depending on the item. Weight ranges should be within those specified.

4. Inspect the trim factors of external fat, length of tail, amount of bone, and so forth.

5. Check the quality and yield grades.

6. Check the condition of the product. Does it possess the attributes of the type and quality product specified? Is product free from discoloration, bruises, blemishes, and so forth?

7. Inspect the packaging material. Is it the type specified? Is it intact? Vacuum-packaged items that have broken seals will not keep as well as intact packages.

8. Check internal temperatures. Products shipped chilled should be below 40°F. Frozen items should be below 0°F and show no signs of thawing.

9. Check extensions of invoice and totals (for example, 12 lb @ $3.00 = $36.00). In some operations this would be an accounting function, not a receiving function.

10. Have the driver initial any discrepancies on the invoice. If the driver does not wait for the product to be checked in, the invoice should not be signed and the right of refusal and credit arrangements should have been previously agreed upon with that supplier. (See cover letter to suppliers in Chapter 14, Purchasing.)

11. The receiving clerk should sign the invoice after completing inspection of the product delivered. This authorization signature is an important control. It is the establishment's assurance that the product was acceptable and that it was delivered. Future problems over billing can be eliminated if an authorized signature appears on all invoices.

15.4 RECEIVING METHODS

There are three basic methods in use today.

INVOICE RECEIVING   Upon delivery, the receiving clerk is given an invoice and follows the steps listed in the preceding section, checking the products against the invoice. This system is the simplest and requires the least time. Its weakness lies in the fact that, if the receiving clerk is in a hurry or lazy, the invoice might be accepted as stated without thorough examination of the product delivered.

BLIND RECEIVING With this system the receiving clerk is given a blank invoice and must fill it out as the product is inspected. The clerk must list all items received and record all their characteristics: weight, grade, condition, and so on. This system forces the receiving clerk to check everything. It is more time-consuming than the invoice receiving method.

PARTIAL INVOICE—PARTIAL BLIND RECEIVING A combined system is sometimes used. With this method, the receiving clerk is given an invoice with the items listed, but the quantity and price omitted. This requires the clerk to weigh each item and record the price. This compromise system offers some of the benefits of the blind receiving method without its time commitment.

The effectiveness of any receiving system can be diminished by unknowledgeable or poorly motivated personnel. It is important that management demonstrate to employees the important role the receiving department plays in maintaining standards and controlling costs within the operation. Appreciation of the importance of the task combined with good training and support programs can help make the receiving clerk a very valuable employee.

Large food service operations with a separate receiving department sometimes use a receiving clerk's report form shown as Table 15.1. This form duplicates much of the information on the invoice, but it also provides a control by designating whether the products have gone directly into production or have been stored or used elsewhere.

TABLE 15.1 RECEIVING REPORT FORM

\#_____

Receiving Clerk's Daily Report

Date_____   Clerk's Name_____

Item	Quantity	Unit	Unit Price	Supplier	Invoice	Total Amount	Distribution		Sundries
							Production	Storage	

Control of the Receiving Department

No matter what method of receiving is used, it is imperative that a person with a high degree of integrity be selected for the job. This is a position of responsibility as important as purchasing, menu planning, or cooking. It is also a job where temptation is present. The receiving clerk is in a position where it is possible to steal merchandise fairly easily or to receive bribes for accepting inferior merchandise or short weights. Unless spot checks are performed by management the receiving clerks can often cover their tracks if they have acted dishonestly.

Separating the purchasing and receiving responsibilities is an important control. If the same person purchases and receives, he is put in a position of temptation. If he is accepting a percentage from a supplier, he may help the supplier recoup that amount or more by purchasing at higher prices, or by accepting inferior merchandise or short weights. With two different individuals filling the positions, there is less likelihood of theft although it is always possible for collusion to occur, though this seems less likely than a single individual's succumbing to temptation. Management must spot check each area frequently.

15.5 WHAT MAY BE DISCOVERED BY INSPECTION

Following the receiving procedures outlined in Section 15.3, the receiving clerk may discover any of a long list of discrepancies between what was ordered and what is delivered. These may include higher prices than on the purchase order; less weight than billed for; excess weight (more than ordered); excess glaze on frozen items like shrimp and crabmeat; repacking items with less count (so that a case of meat that is supposed to have 24 portions has only 20, for example; billing for gross weight including ice or packaging; substitution of merchandise for that specified; mixing in lower quality grades; improper portion sizes; excess trim on product; product that has already begun to deteriorate.

Several options are open to the receiving clerk when discrepancies are found. If the invoice is inaccurate, the clerk can correct the invoice and have the driver initial it and also notify the purchasing agent or management of the discrepancy while the driver is still there, if possible. In case of a minor discrepancy—say, product not trimmed properly—the clerk can send the item back and write a credit memo, or, alternatively, accept the item, noting the discrepancy and notify the purchasing agent to obtain credit from the salesman. In case of a major discrepancy—say lower grade product than ordered or product in poor condition—the clerk should send items back, making note of the action on the invoice and write a credit memo. Notify the purchasing agent and production department of the discrepancy.

The receiving clerk must enforce the purchasing specifications and keep the purchasing agent informed about the purveyors' track records. If there are constant minor discrepancies or periodic major ones in deliveries from a given purveyor, occasionally, entire shipments may have to be sent back, to let the purveyor know you mean business. This policy must be agreed upon by management, and the purchasing agent must know when shipments are refused so that orders can be placed elsewhere to replace the rejected supplies. Another way of discouraging receiving problems is to issue warnings to purveyors, then cut them off from orders for a period of a few months. If these two procedures do not put a supply firm on its best behavior, it is hopeless. Find a new source of supply!

No food service operation can afford to do without an efficient receiving person or department. The discrepancies listed previously are just a sample of problems that proper receiving can short circuit. Not all of those problems come about through dishonesty on the part of the supplier. There are many reputable purveyors who strive to maintain very high standards of integrity in their business relationships. These companies can make honest mistakes, but even honest mistakes are costly if they are not corrected.

Ethical purveyors get stuck with their share of unethical employees. Someone at the plant may be short shipping to make up for products they are taking home. The driver may have a business of his own on the side—he may short weight customers who do not check the product delivered carefully and then sell the merchandise for cash down the road. Some operations compound this problem by allowing deliverymen to carry deliveries directly into the store room. In this situation, the dishonest driver may not only hold back items he is delivering, he may actually steal additional items from the coolers. A competent receiving clerk will prevent unscrupulous deliverymen from engaging in practices that would cost the food service operation additional dollars.

15.6 SUMMARY

It is better to do a hasty job of receiving than none at all. The driver is a direct pipeline to the supplier. Remember that purveyors themselves are receiving product of varying levels of quality and condition. It all must be sold to someone. Often the lower end of the product line winds up in the operations that do little or no checking at delivery time. The goal of purchasing and receiving is to obtain the product most suited to the operation at a reasonable price. This goal can be attained only if the purchasing specifications are enforced at the receiving dock.

The time and effort put into developing a proper receiving program and keeping it effective are well worth the cost. An adequate receiving department is the only way to guarantee consistency in product cost and quality.

CASE STUDY   You are the purchasing agent at the Statler West Hotel, a medium-sized hotel for international tourists. Your menu is continental by design and it offers a great variety of constantly changing luncheon and dinner menu items. Stieger's, a major supplier of meat products in your area, approaches you with the following proposal: "Purchase the #103 primal rib of beef and we will fabricate it for you into a 109 rib, short ribs, stew meat, and ground beef. You are currently purchasing all of these items anyway, and this way you will get the 109 at the price per pound of a 103." This proposal seems logical, but what problems does it present at the receiving dock?

16 | STORAGE

Much can be learned about a food service operation by looking into its storerooms and coolers. Forecasting, purchasing, receiving, and to some extent preparation can be evaluated by a tour of the storage areas. Management's concern and control over all of these critical functions can also be judged by what is seen (or not seen) in the storage department.

Too often food becomes "lost in storage" either because of theft or because of spoilage due to poor inventory rotation or turnover. Millions of pounds of products are lost annually due to evaporation and shrink losses resulting from inadequate storage conditions. In order for a food operation to be successful, it must maintain precise storage conditions and adequate controls over the storage functions.

Functions of the Storage System

1. Protect the product until it is used or processed
2. Serve as a buffer system between purchase and production with back-up product in case a delivery is delayed
3. Provide a safety margin for discrepancies between forecasted use and actual use
4. Make speculation and hence savings possible
5. Maintain or enhance product quality
6. Control inventory to determine the cost of goods sold

16.1 EQUIPMENT, FACILITIES, AND INFORMATION NEEDED

Space and Equipment

The amount of space allocated and the number and type of refrigeration units required will vary with the size and scope of the operation. Almost all food operations require dry goods storage, beverage storage, refrigerator storage (which may be further divided into storage of produce, red meat, fish and shellfish, poultry) and freezer storage.

Information

Reference material is needed to establish guidelines on product keeping quality under various conditions (see guidelines later in this chapter).

Forms

Control of product in inventory is often simplified by the use of various forms, such as storage tags, issuing forms, inventory forms. Although forms are helpful, they themselves do not guarantee control over product. Storage tags are used to identify the type of product they are attached to and the length of time the tagged item has been held in storage. Recording the storage time is particularly critical in the case of high-priced meat items. With modern packaging systems, it is very difficult to judge the length of time a product has been kept because the appearance of the product does not change appreciably, slight changes that do occur generally cannot be seen under the light conditions of the normal walk-in cooler.

The simplest form of tagging is to date stamp the case or container in which the product is stored. Along with this initial date, a warning date might also be applied to signal the time by which the product should be used. Frozen products would have to be used by this date, but fresh products suitable for freezing could be frozen to extend storage life.

A meat **storage tag** like the one shown in Figure 16.1 can be used to identify and date the product. One advantage of using such a tag is that it states the value of the product. A cardboard box containing meat may not be of as much concern to an employee as a box tagged as strip loins and stating the value of the contents at $240. Usually, the storage tag is a two-part form. One part is sent to the person responsible for food cost accounting (the controller) when the product is received. The second part remains on the product until it is issued for use, then the production crew sends the other half to the controller to be cross checked with the first part.

Figure 16.1   Meat storage tag.

#10946 DATE RECEIVED: *9/21*	#10946 DATE RECEIVED: *9/21*
ITEM: *Strip loin* GRADE: *Prime*	ITEM: *Strip loin* GRADE: *Prime*
WT.: *12 lb* @ *$3.00*	WT.: *12 lb* @ *$3.00*
PURVEYOR: *Cross* EXTENSION: *$36.00*	PURVEYOR: *Cross* EXTENSION: *$36.00*
ISSUE BY: *10/6*	ISSUE BY: *10/6*
DATE ISSUED: *10/1*	DATE ISSUED: *10/1*
	ISSUED TO: *Main kitchen*
Part 1. To Controller	*Part 2. Stays with product until it is used.*

For a control, the tag has a number of advantages: it forces the person responsible for receiving to weigh each product. It gives the individual responsible for storage a guideline as to how soon the product must be used and makes stock rotation easier. The tags can be the basis for physical inventories and computing entree food costs on a daily or weekly basis. In multiunit operations the tag tells the controller exactly where that product was issued, for proper cost allocations.

PERPETUAL INVENTORY FORM Larger operations with full-time stewards might utilize a perpetual inventory form like Table 16.1. This form lists items as they are received from various purveyors, their unit and total value, and when and where they are issued. A running balance of product on hand is kept to aid in purchasing and inventory control.

TABLE 16.1. PERPETUAL INVENTORY SHEET

Received						Issued					Balance	
Date	Purveyor	Qty.	Unit	Price	Amount	Issued Date	To	Req. No.	Qty.	Amount	Qty.	Amount

16.2   THINGS TO CONSIDER IN REFRIGERATOR STORAGE

Security

It is good practice to secure all nonproduction-related walk-in coolers and freezers with locks. High-cost items should be protected both against pilferage and against deterioration due to constant temperature fluctuation. As product is requisitioned, it can be transferred to refrigerators and freezers at the point of preparation.

Temperature

Perhaps the most important consideration in the storage of meat, fish, and poultry is temperature. The lower the temperature the longer a meat product will keep, all other conditions being equal. Most bacteria grow at a slower rate as temperature decreases. At 32°F it takes an average microorganism 38 hours to double; at 50 degrees it requires 5 hours, and at 90°F it doubles in 30 minutes.[1] Enzymatic action will also be slowed by lower temperatures. Normal temperature ranges for refrigerated storage would be between 28°F to 38°F. As important as temperature level is the constancy of temperature. If temperature is allowed to fluctuate constantly, the storage time will be reduced.

Table 16.2 shows the drastic effect temperature can have on the keeping quality of fresh cod and haddock fillet.

Quite often because the kitchen's refrigeration units are located close to ovens, broilers, and fryers and are opened and closed constantly throughout the meal period, they are operating close to 45°F. Therefore, only minimum amounts of product should be held in these point-of-preparation units.

Air Circulation

Rapid air circulation is needed to dissipate heat quickly but not so rapid as to cause excess dehydration of the products stored. Even circulation throughout a cooler is important to prevent hot spots where products deteriorate more rapidly than in other areas. Merchandise should be stored off the floor on slats and a few inches away from the walls to allow proper air circulation.

TABLE 16.2   EFFECT OF TEMPERATURE ON KEEPING COD AND HADDOCK

Storage Temperature, °F	Keeping Time
31.5°	12 days
33°	8 days
37°	5 days
45°	2 days
77°	20 hours

Humidity

Probably the most overlooked factor in storage is humidity. Ideally, a cooler's humidity should be at the same level as the level of moisture in the product being stored. If the relative humidity is lower than the moisture content of the product, moisture is extracted from the product and rapid weight loss can occur. The loss can be expensive. For example, a nature-fed leg of veal containing 70% moisture held without packaging in a cooler with a low relative humidity could shrink 5–10%. If the leg weighed 35 lb initially, it could lose 3.5 lb during storage. At $3.50/lb that is a costly storage loss ($12.25).

Sanitation

Sanitation is important in every area of food production and food service, but it is critical in the storage and preparation area. Too often walk-ins and reach-in boxes are neglected. They are not cleaned with the regularity of other parts of the operation.

16.3   PREVENTING BACTERIAL SPOILAGE

Because temperatures are low in storage, we often assume that stored product is safe from spoilage microorganisms. Unfortunately, many bacteria, yeasts and molds, are capable of surviving and multiplying quite readily under refrigeration. Though many of these microorganisms are nonpathogenic (which means that they will not cause food poisoning) they can make meat, fish, and poultry items uneatable. High levels of these organisms can result in severe changes of color, odor, and flavor in the product.

Food Poisoning

Meat is also a potential carrier of pathogenic bacteria. The live animal may be infected or the carcass or finished product can become contaminated during slaughter or processing. Contamination may also occur at the food service operation itself due to poor sanitary conditions or procedures. According to Strange et al., "detectable unfavorable orgenoleptic changes appear when the number of bacteria exceed $10^6/cm^2$ for intact meat."[2] Since pathogenic bacteria not only have the potential to destroy the value of the product, but also have the potential to destroy the reputation of a food service operation should a case of food poisoning occur, they are of major concern to the food operator.

THE INFAMOUS FOUR   The bacteria known as "**the infamous four**" are most often implicated in food poisoning cases; they are
1.   Salmonella
2.   Clostridium perfrigens
3.   Staphylococcus
4.   Clostridium botulinum

Table 16.3 gives information on what products each of the four occur in, how to prevent the bacteria from reaching levels that are harmful and what the symptoms are should an outbreak take place.

TABLE 16.3 BACTERIA THAT CAUSE FOOD POISONING

Salmonella
Occurrence
Raw meats, meat pies, poultry, poultry pies, raw sausage, raw milk, fish, and raw eggs.
To prevent
Use good personal hygiene; avoid excess handling; keep food refrigerated below 45°F; heat food with exposed surface to temperature above 140°F; clean all utensils and surfaces that come in contact with the raw product; and cool leftovers rapidly.
Symptoms of food poisoning
Vomiting, fever, headache, diarrhea, and abdominal discomfort usually appear 24–48 hours after contaminated food is eaten and lasts 2–4 days.

Clostridium perfrigens
Occurrence
High-protein foods, ground meats, food held improperly, and leftovers.
To prevent
Avoid holding food between 70° and 120°F for more than 2 hours; refrigerate food below 45°F; heat food with exposed surface to temperature above 140°F; thoroughly reheat leftover food.
Symptoms of food poisoning
Diarrhea and abdominal pain appear as early as 4 hours after contaminated food is eaten and last up to 22 hours (no vomiting).

Staphylococcus
Occurrence
High-moisture foods, meat, poultry, and egg products.
To prevent
Use good personal hygiene; maintain proper sanitary conditions; refrigerate food below 45°F, handle leftover foods properly. (Food in which Staphylococcus has grown cannot be rendered safe by cooking.)
Symptoms of food poisoning
Diarrhea, vomiting, and abdominal cramps usually appear 2–4 hours after contaminated food is eaten and last up to 24–48 hours.

Clostridium botulinum
Occurrence
Nonacid foods, foods improperly canned.
To prevent
Can food in pressure cookers at appropriate heat levels for the required length of time and boil suspected food 15 minutes before serving to destroy the toxin.
Symptoms of food poisoning
Double vision, difficulty in swallowing, breathing, and speaking, and paralysis of the respiratory system usually appear 12–36 hours after the contaminated food is eaten. Food poisoning is often fatal unless diagnosed rapidly.

Storage Guidelines

To reduce product loss due to spoilage microorganisms follow these guidelines for product in storage:

1. Inspect the condition of the product thoroughly before it is put into storage.
2. Do not keep product beyond the recommended safe storage time.
3. Maintain appropriate temperatures.
4. Minimize handling of product.
5. Do not cross contaminate by storing one product in direct contact with another.
6. Since most bacteria are on the surface of a product, avoid piercing the raw product. (This can cause internal contamination, which might not be exposed to high enough heat during cooking to kill the bacteria.)
7. Store cooked product separate from raw, and use leftovers rapidly.
8. Keep all walk-ins, reach-ins, and equipment that the product comes in contact with (trays, meat hooks, and so on) clean.

16.4   HOW LONG A PRODUCT WILL KEEP

ORIGINAL QUALITY   Generally the higher the quality the longer the product can be stored.

HANDLING FROM SLAUGHTER TO PRODUCTION   The more the product has been handled, the shorter the keeping time. For example, whole chickens keep much longer than chicken parts (legs, thighs, breasts) because during the processing temperatures are elevated slightly and bacterial contamination is increased due to cutting.

SIZE AND SHAPE OF THE CUT   A good guideline is the larger the cut and hence the less exposed surface, the longer it will keep under refrigeration. A strip loin will keep longer than a single strip steak.

FRESH, FROZEN, CURED AND SMOKED, OR COOKED   The order of perishability is as follows:

fresh	shortest shelf life
cooked	
cured	
cured and smoked	
frozen	longest shelf life

Cooking is a method of extending a product's shelf life after it has been held full term fresh. Cooked product has had most surface bacteria destroyed and many enzymes denatured. This is why a top round, for example, can be held full term fresh, then roasted and held in the cooked state for an additional period of time. The major limitation to holding cooked product is the oxidization reaction that occurs producing a noticeable off-flavor sometimes described as a warmed over flavor. Of course, if product is mishandled after cooking, bacteria can again become a factor in spoilage.

Packaging

Whether the product is packaged or not and what type of packaging material it is in helps determine how long it can be stored. There are many systems for packaging wholesale and portioned cuts today. Cryovac, Bivac, and Multivac are three of the major trade names. A product properly packaged usually keeps much longer than an unpackaged product.

SPECIFICATIONS   The following three criteria should be considered when specifying the type of packaging film:
1.   How long will the product be held under refrigeration?
2.   Will it be frozen?
3.   How long will it be held frozen?
If the product is going to be held for a week or more under refrigeration or frozen and stored for more than a month, the packaging film should contain an oxygen and moisture barrier. The oxygen barrier slows down oxidative rancidity and inhibits the growth of aerobic bacteria. The moisture barrier prevents extraction of moisture from the product into the atmosphere of the cooler or freezer.

A film providing both barriers gives the best overall protection. Unfortunately, quite often product is packaged in films that do not contain much of either an oxygen or a moisture barrier. This is done to maintain color. In order to keep the bloomed color (bright red) on beef, an oxygen-permeable film must be used. Without oxygen present, the color would be purple. The appearance of the packaged product is more of a concern at retail than in a food service operation. Knowledgeable buyers and cooks understand the reason for the purple color and know that after the product is removed from the vacuum-sealed package, the color will bloom.

The importance of packaging and in particular the moisture barrier is shown by the following comparison of purge losses for different amounts of storage time for wholesale cuts packaged in plastic film.[3]

Number of days	3	7	14	28
Purge loss, %	0.5	0.6	0.7	0.8

Wholesale cuts held unpackaged for 28 days would suffer not only moisture losses, but also trim losses due to surface drying and mold growth that would have to be cut away. It would not be unusual to experience more than 10 percent loss in this unpackaged product.

Some products such as frozen shellfish are not always packaged in film. Often they are protected by an application of water referred to as a glaze. For example, when a 5-lb block of shrimp is frozen, it is also coated with water to form a ⅛ in. to ¼ in. ice cover. This cover helps prevent extraction of moisture from the shrimp while it is held in frozen storage. Over a period of time, sublimation gradually removes the glaze and exposes the shrimp. The shrimp should be used before this happens or the product quality will suffer. **Note:** The glaze can be reapplied by dipping the frozen block of shrimp in water in the freezer; repeat several times until surface ice has been built up.

Table 16.4 is based on the foregoing considerations to provide some guidelines for the average good-quality storage life of meat, fish, and poultry under refrigeration. This information can be used as a starting

TABLE 16.4   AVERAGE GOOD-QUALITY STORAGE LIFE OF REFRIGERATED MEAT, FISH, AND POULTRY

Type of Meat	Temperature, °F.	Time Limit	Relative Humidity,%*	Special Considerations
Beef				
Major wholesale cuts	34–36°	1 week	85	Unwrapped
Major wholesale cuts	34–36°	2 weeks		Cryovaced (barrier film)
Portion beef cuts				
New York strip steak	34–36°	4 days	85	Layer packed
New York strip steak	34–36°	1–2 weeks		Bivac or Multivac (barrier)
Ground beef	34–36°	3 days	85	Cellophane or Saran wrap
Pork				
Wholesale cuts (fresh ham, loin)	34–36°	5 days	85	Unwrapped†
Portion Cuts (chops, cubes,	34–36°	3 days	85	Layer packed
pork sausage)	34–36°	3 days	85	Cellophane wrapped
Chickens, whole	28°	10 days	95–100	Ice packed
	34–36°	6 days	85	Unwrapped
Lamb, wholesale cuts				
(loins, legs)	34–36°	1 week	85	Unwrapped
Portion-cut lamb	34–36°	4 days	85	Unwrapped
	34–36°	1–2 weeks		Bivac or Multivac (barrier)
Veal, wholesale cut				
(leg, loin)	34–36°	5 days	90	Unwrapped
Veal, portion-control cuts	34–36°	3 days	90	Layer packed
Fresh fish	28°	2–5 days‡	95–100	Crushed ice
Shellfish				
Clams and Oysters	34–36°	1 week	85	No ice or water
Cured meats, ham, bacon,				
corned beef	34–36°	2 weeks	75	Cellophane or Saran wrap
Variety meats	34–36°	3–5 days	85	Layer packed or Cellophane wrapped

*Humidity is not a consideration for products packaged in a moisture barrier film.
† Not much pork is available in Cryovac.
‡ Varies with type of fish, amount of processing, and distance from source of supply.

point for establishing the time limits for holding different products in refrigerated storage. The number of days listed are the average keeping times after the product has been received into a food service operation. Most products, with the exception of ground beef and fresh sausage, have had approximately 7 days of refrigerated storage prior to delivery. Actual shelf life will vary from one operation to the next depending on the condition of the product on delivery and other factors discussed in this chapter. (Table 16.5 in the next section presents guidelines for keeping products in the freezer.)

16.5   FREEZING VERSUS FROZEN STORAGE

Most food service operations are equipped with low-temperature refrigeration units designed to hold already frozen product, but not designed to freeze the product. These units operate in the temperature range -10°F to 0°F. Although it is possible to freeze product at these temperatures, the process is slow and hence the products' quality is reduced.

Rapid Freezing versus Slow Freezing—A Palatability Difference

Today commercial freezing is done by blast freezing at -40° to -60°F or by use of cryogenic freezing tunnels using liquid nitrogen or liquid CO^2 at extremely low temperatures (-290°F). These systems freeze the product rapidly and by doing so produce very fine ice crystals within the cell which do not appreciably damage the cell structure of the meat product.

When high temperatures are used and the freezing rate is slow, large extracellular ice crystals are formed, which may rupture the cell membranes. Once the cell membrane is damaged, additional protein denaturation occurs, and as a result there is a loss in the water-holding capacity of the muscle. Upon thawing, therefore, the product suffers a significant increase in drip loss compared to a product frozen at an adequate freezing rate. The cell damage and concomitant moisture loss has a detrimental effect on the texture and juiciness of the end product.

If a Product Must Be Frozen on the Premises at 0°F

Although ideally freezing should be done using lower temperatures than 0°F, in reality food service operators are often faced with no alternative. If a strip loin valued at $45.00 cannot be used within the safe refrigerated shelf life, it must be frozen on the premises. Some palatability may be lost, but the entire item will be lost if not frozen, so there is really no choice. To minimize palatability losses, follow these guidelines:
1.   Freeze the product while it is still in good condition.
2.   Package properly in vacuum type film, which has a moisture and oxygen barrier or hand wrap in a barrier film expelling as much air as possible.
3.   Freeze in the lowest temperature unit available.

4. Get the fastest freezing rate possible by freezing single layers. Do not freeze an entire case or cases of strip loins, for example. Instead, lay each strip loin out on a shelf or pan until frozen, then repack in case lots. This reduces the penetration factor and significantly increases the freezing rate.

5. Use the frozen product as soon as possible.

Criteria for Estimating Keeping Quality

PRODUCT QUALITY AND CONDITION AT THE TIME OF FREEZING It is impossible for a USDA Good strip steak to come out of the freezer with the palatability characteristics of a USDA Prime. It is also not logical to freeze an item that has been held so long under refrigerated storage that it is stale, expecting to resurrect it by putting it in a freezer for two or three months. The quality of the product and the amount of handling it has had prior to freezing will have a dramatic effect on the shelf life of the frozen product.

PRODUCT TYPE The structural and chemical composition of a product affect the product's keeping quality in the frozen state. Beef in general will keep longer than pork. The major reason is the makeup of the fats and their susceptibility to oxidization; pork fat becomes rancid much more quickly than beef fat. Because of varying oil content fish of different species also vary significantly in the length of time they will keep. Certain species of fish exhibit a toughening of the muscle when frozen that is believed to be caused by protein denaturation during freezing and with subsequent storage.

PACKAGING MATERIAL (see discussion, Packaging, in Sec. 16.4.)

FREEZER TEMPERATURES To maintain quality of frozen items, storage temperature should be 0°F or below. The lower the temperature, the longer the optimal storage life of the product. As important as low temperature is constant temperature. If temperature is allowed to fluctuate by 10°F or 15°F, two adverse effects may occur. The first is increase in the size of ice crystals growing inside the product. A certain amount of free water becomes available as the temperature rises; upon freezing, this water attaches to existing crystals, forming larger crystals, which may cause additional drip loss.

The second adverse effect is sublimation, whereby a portion of the solid water is removed from the product in the gaseous state when the temperature rises. This gas then condenses on the product to form a snowlike cover, but the moisture is lost from within the product forever. These two effects can substantially decrease palatability of the product.

Table 16.5 lists average good-quality storage times for various meat, fish, and poultry products stored frozen. Actual keeping time will differ from one operation to the next because of the many variables.

TABLE 16.5 AVERAGE GOOD-QUALITY STORAGE LIFE OF FROZEN MEAT, FISH, AND POULTRY, IN MONTHS

	0°F	−10°F	−20°F
Beef			
Wholesale cuts	6–8	10–12	14–16
Portion cuts	4–6	8–10	10–12
Ground beef	4	6	8
Lamb			
Wholesale cuts	6–8	10–12	14–16
Portion cuts	4–6	8–10	10–12
Pork			
Wholesale cuts	4–6	6–8	8–10
Portion cuts	2–4	4–6	6–8
Cured and smoked pork	1–2	1–2	1–2
Veal			
Wholesale cuts	4–6	6–8	8–10
Portion cuts	2–4	4–6	6–8
Fish			
Fatty type	2–3	4–6	6–8
Lean fish	3–4	6–8	8–10
Shellfish			
Shrimp and scallops	3–4	6–8	8–10
Clams and oysters	2–3	4–6	6–8

Recommended Methods for Thawing Frozen Product

UNDER REFRIGERATION (32–36°F) Refrigeration is generally considered to be the best and safest way to thaw frozen product. It requires accurate forecasting because large items—say, a 22-lb 109 rib of beef—will take 3 to 4 days to thaw totally whereas smaller items thaw more quickly (a 12-oz portion-cut strip steak would thaw in approximately 12 hours).

SUBMERGED IN COLD RUNNING WATER If the product is packaged in a moisture-proof film, it can be submerged in cold running water to hasten the thawing process. The cold water keeps the surface temperature from rising to dangerous levels. A 12-oz portion-cut steak can be thawed by this method in approximately 30 minutes.

RAPID THAW IN A CONTROLLED ENVIRONMENT Use a low-temperature oven setting (200°F) to rapidly thaw large roasts in emergencies. The 22 lb 109 rib that takes 3–4 days to thaw under refrigeration can be successfully thawed in approximately 3 hours in a slow oven. Rapid thaw is safe if the surface temperature of the meat is exposed to 200°F oven temperature because the heat eliminates the dangers of surface bacterial growth. Once the rib is thawed, the oven temperature can be adjusted to normal roasting temperature.

MICROWAVE The microwave can be used for thawing some individual portions. Care should be taken to pulsate the amount of energy applied to the product while thawing: shoot for 30 seconds, let the product rest for 1 minute, then shoot again for 30 seconds, then rest, repeating until product is completely thawed. If the energy is not applied in pulses but continuously, the surface moisture becomes thawed too quickly and will be boiled away before the interior thaws.

A DON'T Do not thaw at room temperature—the risks are too great!

16.6   SUMMARY

Storage is certainly a key subsystem of a food service facility. Management must set up storage guidelines and policies that are realistic and workable for their operation. But management should not stop there: unscheduled periodic visits to the coolers, freezers, and storeroom areas should be made so that constant reevaluation of this critical area is possible. Product "lost in storage" will never generate a profit on the income statement!

CASE STUDY

You have been hired by SAM's Catering Service to set up storage guidelines. This high-volume service purchases both bulk orders to be held frozen for three to six months and weekly orders held fresh and used according to need. The following lists show the average weekly fresh order and a typical six-month bulk order:

Weekly Order	Bulk Purchase (3–6 months)
200 lb tenderloin	10,000 lb Choice 109 ribs
300 lb chickens	5,000 lb green headless shrimp
5 bushels cherrystone clams	20,000 lb turkeys
50 lb fresh fillet of sole	1,000 lb King crab legs
50 lb racks of lamb	1,200 lb veal legs
100 lb pork chops, center cut with pocket	
60 lb bacon, hotel pack	
200 lb ground beef	

Adequate walk-in freezers and coolers are available, but no well-defined storage policies have been set. It is your job to design a control system for this inventory and to specify exact storage conditions for these products. Include the packaging requirements, temperatures, humidity, and recommended safe time limit for storing each item. Consider thawing methods and timetables for frozen product.

REFERENCES

1. D. B. Devendorf, Improving the Shelf Life of Meat Products, *Meat Processing* (May 1976).

2. E.D. Strange, R.C. Benedict, J.L. Smith, and C.E. Swift, "Evaluation of Rapid Tests for Monitoring Alterations in Meat Quality during Storage." *Journal of Food Protection,* vol. 40, no. 12, pp. 843–847, Dec. 1977.

3. From a study in *Journal of Food Science* 143 (January-February 1974).

17 | COOKING METHODOLOGY

A tremendous amount of misinformation is published in cookbooks concerning food preparation methods. Bad techniques have been passed down from one generation of cooks to the next and little scientific knowledge has been applied to the so-called art of cookery. The intent of this chapter is to dispel some of the cooking myths and give the fundamentals of some improved approaches to the preparation of entree foods.

Objectives of Cooking Meat, Fish, and Poultry

In preparing entrees commercially, the primary objective should be to increase or at least maintain the palatability of the main ingredient of the entree—the meat, fish, or poultry. This means bringing out or adding to the tenderness, flavor, and juiciness of the product. Cooking should enhance the appearance of the product by bringing it to the right color on the outside while it reaches the desired degree of doneness inside. When these objectives are achieved, the customer receives an entree cooked to perfection. Management has of course one additional objective, and that is to prepare the product with the least shrink loss and using the least energy.

Heat Transfer—A New Approach to Explaining Cooking Methods

Most cookbooks list and define nearly a dozen different methods of preparation. Paragraphs, if not whole pages, may be used to explain each method. Table 17.1 represents a different approach to explaining cooking methodology. This approach focuses on the heat transfer mechanisms in cooking —on the way heat is conveyed or passed from the heat medium or object to the food. The primary mechanisms are convection and conduction or a combination of the two. **Conduction** is the transmission of heat from one surface directly to another, as when heat is transferred from the surface of a hot grill to the surface of a steak. **Convection** is the transfer of heat by the circulatory motion of fluids or gases that are heated. Heat is transferred to a roast by the movement of heated air in an oven. But note that convection brings the heat to the surface of the roast, then heat is **conducted** to the interior. This approach calls attention to the medium

251

Table 17.1 COOKING METHODS AND HEAT TRANSFER*

Method	Heat Transfer Mechanism(s)	Heat Transfer Medium	Amount and State of the Medium	Temperature of the Medium, °F	Temperature Curve
Roasting	Convection, conduction†	Air		165–400	Constant
Baking	” ”	”		325–475	”
Broiling	” ”	”		350–700	”
Pan broiling	Conduction	Pan direct		250–450	”
Grilling	”	Grill surface		250–450	”
Sauté	Conduction, convection	Oil, fat, butter	Small amount, partially immersed	250–375	”
Deep fat frying	” ”	Oil or fat	Fully immersed	325–375	”
Boiling	” ”	Water	” ”	212	”
Simmering	” ”	Water or stock	Immersed	Not allowed to boil, 200–210	”
Poaching	” ”	” ”	Minimum amount of liquid	165–200	Not constant
Braising	” ”	Fat then water	Immersed	195–212	” ”
Pot roasting	” ”	Water or stock	”	On stove, 180–212; in oven, 325–375	” ”
Steaming	” ”	Air, water vapor	Pressurized	Depends on pressure	” ”
Stewing	” ”	Water or stock	Just to cover	190–212	” ”
Microwave	None No heat transferred				

* Heat transfer is the passing of heat from one medium or object to another.

† Convection is the transfer of heat by the circulatory motion of a fluid (water or air). Conduction is direct transmission of heat from one surface to another.

of heat transfer (oil, water, air) and the temperature of the heat transfer medium and fluctuation in the temperature. All conventional methods of preparation can be represented in terms of heat transfer. The only exception is microwave cooking, which has no direct heat transfer but, rather, an energy transfer that produces heat. The best cooking method for a particular entree depends on a number of factors: the cut to be prepared, its quality, size, and shape, and the end product expected by the customer. (Specific recommendations were made under **Utilization** in the product chapters.)

Changes during Cooking

COLOR Surface browning is caused by the reaction of amines with reducing sugars. For beef, for example, this reaction begins when surface temperature reaches approximately 194°F. Interior color changes are due to denaturation of the pigment myoglobin. At different temperatures, varying amounts of myoglobin are denatured producing characteristic colors—the red associated with rare (120°F), the pink with medium (140°F) and the gray with well done (160°F). (See color insert p. 32.)

REDUCTION IN WATER-HOLDING CAPACITY Cooking causes an alteration of muscle proteins which results in the loss of moisture in the form of drip and evaporation losses. As the level of heat and the internal temperature reached increases, shrinkage increases.

MUSCLE SHORTENING Muscle proteins go through two stages of shortening, the first between 104°F and 122°F, the second between 149°F and 167°F. The result of muscle shortening is a toughening of the cooked product. The second stage of shortening, sometimes called protein hardening, is much more detrimental to tenderness than the first stage.

CONNECTIVE TISSUE CHANGES Upon heating, collagen first begins to shrink at approximately 135°F; above 140°F it begins to form gelatin.[1] These two changes in collagen can significantly increase the tenderness of a cooked product.

FLAVOR AND AROMA Various volatile constituents are dispersed during cooking that contribute to the characteristic odor and flavor of cooked meat. The intensity of these flavors and aromas varies with the type of meat, the level of heat used in cooking, and the internal temperature reached.

The Number 1 Palatability Factor: Tenderness

All of the changes produced by cooking are important because they affect the palatability of the finished product. The two factors that affect tenderness are emphasized here because the product's tenderness influences the diner's judgment of overall palatability so much.

When cooking a cut of meat, a tenderness trade-off must be considered. As the muscle protein cooks, it becomes tougher. The higher the heat it is subjected to, the higher the internal temperature of the meat, and consequently the less tender it will be. The reverse is true concerning one of the connective tissue proteins, collagen. When exposed to heat for extended periods of time and allowed to reach a temperature of 140°F or higher, collagen will first shorten, which causes a slight increase in tenderness, and then it will hydrolize and form gelatin, which significantly improves tenderness.

This trade-off should be kept in mind when selecting the best cooking method and determining the internal temperature for a cut of a particular nature and amount of connective tissue. A center cut strip loin steak has very little connective tissue and should be broiled relatively quickly and to a rare degree of doneness (120°F) to avoid unnecessary muscle toughening. A nerve or vein end strip steak, however, contains a thick piece of collagen running through the center of the steak. Consequently it must be broiled slower and to a medium (140°F) or above degree of doneness to ensure the softening of the connective tissue. If the second stage of muscle shortening (149°–167°F) is avoided, the steak still has a very acceptable level of tenderness.

In comparison, chuck steak, tougher because of a much higher percentage of collagen than the nerve end strip, might be cooked by braising (using moist heat) for a long time so that it reaches an internal temperature well above 149°F. After prolonged braising the muscle of the chuck steak would be tender but would exhibit a different sort of tenderness than that of a tenderer steak cooked rare. The tender rare steak would be soft and juicy; the well done braised steak would be relatively dry, and, although tender, would have lost integrity so that when chewed, it would crumble.

17.1 METHODS OF PREPARATION

Because the majority of meat items prepared in the food service industry are either roasted or broiled, these two methods of cooking will be discussed in detail. Grilling, barbecuing, and pan broiling are similar to broiling, and the concepts discussed under broiling also apply to these methods of preparation. **Grilling** is conduction of heat directly from the grill to the product. **Pan broiling** is conduction of heat directly from a heavy frying pan or skillet to the product. **Broiling** is cooking by convection of air beneath a heat source. **Barbecuing** is cooking over charcoal or wood; conduction and convection serve as the heat transfer mechanisms.

Moist heat preparation, whether it be stewing, poaching, pot roasting, or braising, all involve the addition of liquid and rather long-term cooking. Because a subjective evaluation of tenderness is generally used to judge when the product is finished rather than the degree of doneness (temperature), any written description of these methods would be less precise than the information supplied in Table 17.1.

The use of the microwave oven for initial preparation of entrees in food service is very limited. It is utilized more often to thaw cooked product, to reheat it, or to finish it to a higher degree of doneness after conventional cooking.

Roasting—Low-Temperature, Even-Heat Method

Whether roasting a product in a conventional oven, a convection oven (using forced air circulation) or one of the new low-temperature cookers (a well-insulated oven that operates at very low temperatures), the objectives are the same: to produce the proper amount of color development on the exterior of the product while obtaining the desired degree of interior doneness, maximizing product yield, and minimizing energy cost. This method, which might be termed the even-heat method, produces the best results from the standpoint of palatability. There are many benefits from low-temperature roasting. First, the interior temperature of the product is kept relatively high—60°, 70°, or 80°F— for an extended period of time. This creates a condition very favorable for accelerated enzymatic activity, which results in an extremely tender product. Second, the product maintains its natural juices and is consequently more flavorful with much better yields. The roast prepared with slow, even heat can also be much more easily carved than one subjected to a higher heat treatment and bearing heavy crusting on the surface as a consequence.

The old theory that initial searing of the product exterior seals in the juices has been proven false many times over. Searing actually dries out the exterior and produces a product with significantly lower yield that is drier and less tender. Many studies have been done showing the greater yield from slow oven roasting. In one such study a dramatic difference in yield was found: Number 109 ribs were cooked in two batches to an internal temperature of 137°F. One batch was done in ovens at 250°F and another batch in ovens at 450°F. The average weight loss at 250°F was 9.8%, whereas the average loss at 450°F was 29.1%. The same test was repeated with fresh hams; the hams roasted at 250°F lost 13% on the average, compared to 30% for the hams cooked at 450°F (both batches were cooked to an internal temperature of 170°F).

HOW TO TELL WHEN A ROAST IS DONE   The best way to determine degree of doneness is by using a meat thermometer to check the internal temperature of a roast. Timetables may be useful guidelines, but no two stoves are alike in the exact amount of heat they produce and controls are rarely accurate enough that the temperature in the oven exactly matches the number you have set on the control. The quantity of product cooked at one time also affects the cooking time. Whether the product is frozen hard, partially frozen, thawed, or at room temperature also alters cooking times. Checking the internal temperature of the roast is the only accurate method to judge whether it is done. The internal temperatures shown in Table 17.2 have been widely used by many food service operators. They may require adjustments depending on regional preferences of the consumers.

ALLOWING THE ROAST TO STAND   When the roast reaches the internal temperature listed in Table 17.2 it should be removed from the oven. The meat does not stop cooking when taken from the oven because there is a temperature differential between the outside and the inside of the roast—the outside temperature is higher, having been directly exposed to the oven temperature. The internal temperature of the roast may rise anywhere from 5° to 10°F after it has been removed from the oven. It is very important to allow the roast to stand prior to carving. This permits the exterior and interior temperatures to equalize and the

TABLE 17.2   INTERNAL TEMPERATURES OF ROAST MEATS, °F

	Beef	Lamb	Veal	Pork	Poultry
Rare	110–120	125–130			
Medium	130–140	135–150	135–150	160	165
Well	150–160	150–160	150–160		

muscle to retain the natural juices. It is also easier to carve a roast that has been given ample standing time, because the roast becomes firmer and therefore much easier to slice either by hand or by machine. A large roast, 18 pounds and over, requires a minimum of one-half hour to one hour standing time.

Broiling

The difference between broiling and roasting is basically the intensity of the heat and the size of the product being prepared. Commercial broilers, whether they be charcoal-fueled with the flame below or the conventional broiler with the flame above, operate at relatively high temperatures (400°–700°F). Often the temperature is too high for the product being prepared and scorches the surface, producing a drier, less tender product than could be obtained with a slower, more even broiling technique. The objective is the same as in roasting: to produce the desired interior degree of doneness at the precise moment that the proper exterior color has developed. The desired degree of doneness and the thickness of the steak therefore must be taken into consideration. A thin steak would be prepared under relatively high heat whereas a thick steak would be prepared with less heat intensity. A thick steak to be cooked well-done should be started at a lower temperature than a thick steak cooked rare. This can be accomplished by either reducing the heat of the broiler or increasing the steak's distance from the heat source.

HOW TO TELL WHEN STEAKS ARE DONE Internal temperatures are the most accurate way to determine degree of doneness when practical. The temperatures listed for roasting may also be used for steaks. It is difficult to check the internal temperature of a steak thinner than ¾ inch, but the degree of doneness can be judged by the feel of the muscle. An indentation will remain in a raw steak where it has been touched. As the steak cooks, the proteins firm up. On a rare steak the indentation springs back slowly; the indentation produced by touching a medium steak springs back rapidly; and a well-done steak has a firm muscle that shows almost no indentation upon being touched. Remember, a steak does not stop cooking immediately after it is taken from the broiler. It takes three to five minutes for the heat to equalize within the steak. During this time, the internal temperature continues to rise. For this reason, it is suggested that cooks pull steaks off the broiler at a degree of doneness slightly under what is ordered. In this way, if there is any delay in pick-up or if the guest is finishing a first course, the steak will not be overdone. More fire can always be put on the product, but a medium steak can never be made rare.

BROILING FROZEN STEAKS The difference between cooking meat from the frozen state rather than from the thawed state is simply one of time and temperature. When a product is cooked from the frozen state, rapid thawing is accomplished first and then the product is cooked to the desired degree of doneness. Cooking from the frozen state requires a longer period of cooking and may also require slightly lower initial temperatures to prevent overcooking the exterior of the product, but it can yield a very appetizing finished product. All steaks can be successfully prepared from the frozen state, although it may not be practical for a very thick steak (over 1 inch thick) simply because the total cooking time may be too long (25 minutes). Thinner steaks are sometimes easier to prepare directly from the frozen state than from the thawed state, particularly if the desired degree of doneness is rare to medium rare. For example, a sandwich steak, weighing 4 ounces and ¼ inch thick, is difficult to cook rare from the thawed state because the conduction of heat occurs so rapidly. The inside of the product is almost cooked to medium before the exterior has had a chance to develop a proper color. But in the frozen steak heat transfer is slower, and as a result a rare steak with good exterior color is possible. The broiler setting must be adjusted to the thickness of the steak. For example, steaks up to ½ inch thick can be broiled under medium to high heat. Steaks ½ inch to ¾ inch thick would require medium heat. Steaks over ¾ inch thick should be started on low heat and then turned up to medium to finish the cooking process.

17.2 COOKING IS A SERVICE

Cooking is just one of the services provided by the food service industry. To begin with, the consumer cannot generally purchase meat of the high quality available to the restaurant or hotel (it is rare to find Prime at retail). In addition, meat purchased retail is generally not aged or tenderized in any way. The overall palatability of meat available in supermarkets and other retail stores might be judged inferior to that available to the food service industry. Moreover, the equipment and facilities available commercially are far superior to home units and the technical knowledge of the preparation staff should enable them to cook meat with a consistently high level of tenderness, juiciness, flavor, serving it at the level of doneness the customer requests.

One problem in providing satisfactorily cooked meat is that the terminology used to describe the degree of doneness is not always a standard. The consumer asks for rare, medium, or well-done meat, but individual perceptions of these degrees of doneness may differ from the commercially accepted ones. Often consumers ask for a steak cooked pink or red to pink. These terms allow a wide margin of error. Because of differences of interpretation and the fact that the product does not stop cooking once it is removed from the broiler or oven, it is safer to prepare the product slightly under the degree of doneness ordered by the customer. The cooking process can always be continued, but it can never be reversed.

Roasts present a problem in satisfying the customer's desire for a particular degree of doneness. It is not practical to roast expensive cuts of meat such as prime ribs or roast strip loin to different degrees of doneness, since forecasting the number of rare, medium, and well-done orders on any given night is virtually impossible. The standard practice therefore is to cook all roasts to the rare degree of doneness, using the end cuts for the more well-done portions and adding heat to the individual portions to produce medium-rare, medium, and so on. There are a number of ways to accomplish this effectively. If the customer wants his order medium-rare and the roast is cooked rare, slice the individual cut and pour the natural juices over it to continue the cooking process. This works well with degrees of doneness up to about medium. If someone orders the cut medium well to well, place the rare cut on a sizzle platter or oven pan with a small amount of the natural juice. Place the pan in a relatively hot oven or salamander section (top oven) of the broiler. Only a few minutes are required to bring the individual slice to medium well or well done.

17.3 A DIFFERENT APPROACH TO ROASTING – THE WATER BATH

Conventional roasting has the disadvantage of causing shrink loss in excess of 15 percent. In order to reduce shrink loss, it is necessary to roast at lower temperatures than conventional ovens permit; most ovens do not operate accurately below 225°F. Some of the newly designed ovens that operate at 165°F reduce shrink below 12 percent, but another, simpler method of cooking meat by dry heat, that is, roasting, is possible. Instead of relying on air for the heat transfer, the product is placed in a roasting bag and a water bath is used for heat transfer. The advantage to this system is that the water temperature can be maintained accurately at lower levels than possible in most ovens. A cut like a top round of beef, for example, placed in an oven bag, sealed, and then submerged in a water bath could be dry roasted with shrink loss below 10 percent. The water does not come in contact with the meat. It is simply used to transfer heat through the bag to cook the product. Surface browning is obtained by using a caramelization agent on the surface of the product if it is desired. A number of advantages that can accrue from this system of cooking:

1. Since the roast is placed inside a bag and cooked in a water bath, there is absolutely no chance for any evaporation loss. Therefore, any shrink or drip that occurs from the roast during the cooking process is captured in the bag and can be used for making a suitable sauce or gravy or serving *au jus*. In conventional oven roasting, there is a certain amount of evaporation into the atmosphere, causing these juices to be lost.

2. The differential in temperature from the exterior to the interior is slight. Consequently the roast is uniformly done from top to bottom with very little difference in color or degree of doneness. In conventional roasting at temperatures of 250°F or higher, the temperature

differential between the exterior surface and the center may be as high as 130°F. The center of the roast is rare at 120°F, but toward the exterior the roast is more well done, since the surface temperature is 250°F. Since the water bath is at 165°F there is only a 45° differential (165° − 120°); therefore the roast is more uniformly done throughout.

3. Because the cooking temperature is maintained at a very low level and the cooking time is extended over a longer period, the naturally occurring enzymes in the meat have much more time to tenderize the meat during the cooking process. This is actually a method of "accelerated aging" because it is increasing the internal temperature of the roast under controlled conditions while maintaining a high enough external temperature to prevent the growth of surface bacteria. It is possible to bring less tender, lower grade cuts to a very acceptable level of palatability; one may try substituting bottom rounds for tops or using ungraded products rather than products graded Choice.

4. Because the bag is sealed, the oxygen is used up very rapidly during the cooking process. Hence the formation of warmed-over flavors is inhibited. Products prepared in this manner, held under refrigeration in the sealed bag and then reheated, do not suffer the off-flavors and odors that normally occur in reheated roast beef products.

5. Because the surface of the product is brought to a temperature above 145°F, most surface bacteria should be destroyed. This can improve cooked storage life (in fact some commercially prepared product is being advertised as having three to six weeks unfrozen cooked shelf life). The authors are not suggesting to keep the product for this period of time, but merely pointing out another advantage of this particular system.

Note: Roasts that have been pierced by blade tenderizing should not be prepared by this method since internal contamination may occur during piercing.

Procedure for Cooking Beef in a Bag in a Water Bath

Note: For this technique it is important to use only heat-stable bags especially designed for cooking meat—for example, Mylar roasting bags. Plastic bags that are not heat stable can permit the transfer of toxic plastic monomers to the food product.

STEP 1 Obtain the accurate net weight of product to be roasted by removing all packaging, weighing the beef, and recording the weight.

STEP 2 If normal oven-roasted color is desired on the exterior of the product, it is necessary to coat the product with a caramelization agent prior to putting it in the bag. Seasoning companies sell beef rubs composed of spices, hydrolized vegetable protein, beef base, and other ingredients that give a typical roasted appearance to the product and imparts some additional flavor to it. If no flavor additives are desired, liquid caramel color can be rubbed over the roast prior to inserting it into the bag.

STEP 3 Insert a meat thermometer into the center of at least one of the roasts to be prepared. If all of the roasts are the same weight range, one or two thermometers within an entire batch should be sufficient to judge internal degree of doneness.

STEP 4 Hydrate unflavored gelatin with the proper amount of water. Weigh each roast and add gelatin solution amounting to 2½ percent of the weight of the roast. Simply pour it into the roasting bag.

STEP 5 Insert the roast that has been treated with caramel color and has the thermometer placed in it, into the bag. If a vacuum machine is available, draw a vacuum on it. If there is no vacuum machine available, submerge the roast in a water bath, draw all the air toward the top of the bag, and seal tightly by either clipping or tying with a string.

STEP 6 Submerge roasts in a water bath that has a temperature maintained at 165°F or above. A suitable piece of equipment for an individual roast is a stock pot if one can maintain adequate temperature control within it. A very useful piece of equipment for quantity production would be a large steam-jacketed kettle (a trunnion) equipped with a solenoid valve to control the steam flow and maintain exact temperature. (In a large trunnion, it is possible to do 15 or 20 roasts at a time.) The lower the temperature, generally speaking, the more yield will be obtained. There is obviously a trade-off between the time it takes to cook the product and the amount of shrink acceptable. It is important to maintain at least a 165°F water bath temperature to be sure that surface bacteria are not allowed to grow.

STEP 7 Cook the roasts to the internal temperature desired. Keep in mind that the temperature rise when the roast stands after cooking will be significantly less than for a normal, oven-roasted product. This is due to the fact that there is such a low temperature differential from the exterior of the roast to the interior. In oven roasting at 300°F, the internal temperature may be brought to 120°F prior to removing from the oven and within ½ hour the internal temperature will increase anywhere from 5° to 10°. With the water bath method, since the internal temperature varies only 40° from the external surface temperature, the set-up will be minimal. Therefore, one may cook to a higher internal temperature prior to removing from the water bath. Some temperatures for various degrees of doneness are 130°F for rare, 140°F for medium, and 150°–160°F for well done.

When the roast has reached the internal temperature desired, if it is to be held under refrigeration prior to service, the best method is to drain the hot water from the kettle, cover the roast with cold ice water and chill it very rapidly. Bring the internal temperature down to about 50° prior to placing the roast in refrigeration. If the roast is to be used immediately, simply cut the bag, drain the liquid into a suitable container (this may be used for natural juice or beef gravy), weigh the roast and calculate the cooking losses. Check temperatures and visually inspect the degree of doneness to see if it suits your particular clientele.

Caution: When cooking beef in a bag in a water bath, all of the oxygen is used up quite rapidly. This creates a situation that could provide an opportunity for Clostridium botulinum to grow, should there be any present on the meat product. Food poisoning from this cooking process is highly unlikely, however. It is possible, but only under conditions where the product has been grossly mishandled. To safeguard against this possibility, the product should either be rapidly chilled after cooking or removed from the bag and used within a reasonable length of time. The roasts should not be left in the bags at room temperature for long periods of time after the cooking process. Once again, the best course is to ice the bags to reduce the internal temperature of the meat to 45°F, then refrigerate the product immediately if it is to be used at some future date.

A study done by Buck et al. in 1979[2] compared conventional low-temperature roasting with the water bath method. A taste panel rated the water bath samples higher in all palatability aspects than the oven-cooked samples. The water bath samples had significantly higher yield, more uniform appearance, and increased tenderness when compared to the oven-cooked samples.

The study by Buck et al. did not directly investigate the microbiological aspects of water bath cooking, but it did review the literature in this regard. Two studies were cited, one investigating Salmonella[3] and one concerned with C. perfringens.[4] The first study showed that beef contaminated with Salmonella cooked to 51.67°C (124°F) still showed low levels of bacteria surviving. Beef cooked to 57.22°C (135°F) and held for three minutes were free of Salmonella.

The study on C. perfringens showed that as internal temperatures reached 51°C (124°F) growth was inhibited and at 55°C (131°F) inactivation of C. perfringens began. Buck et al. postulated from their heating curve data that water bath cookery might have some bacteriological advantages over low-temperature oven roasting. Specifically, the water bath method offers faster rates of heat penetration, which quickly bring the product through the 23.9–51.7°C zone, which is the zone of most rapid bacterial growth. The moist surface of water bath cooked meat may also be beneficial. The surface moisture causes increased heat sensitivity of any bacteria that may be present.

Because safety from food poisoning is of concern, the food service operator should read such microbiological studies. Currently the USDA has guidelines for safe commercial production of cooked roast beef. The recommendations are listed in the following section. Cooks interested in using the water bath method could follow these temperature/time guidelines for added safety.

Statistical Data

Yield tests were performed using the water bath technique, and the results are provided to assist the reader in trying this method. The tests were done in steam-jacketed kettles using split top rounds of beef averaging 10 lb apiece. Table 17.3 is the record of water temperature and internal temperature of the product.

TABLE 17.3   TEMPERATURE CHART FROM WATER BATH COOKING TEST

Time	Water Temperature	Internal Temperature, °F
12:00 Noon	165°F	40°
1:00 P.M.	"	52°
2:00 P.M.	"	70°
3:00 P.M.	"	95°
4:00 P.M.	"	112°
4:45 P.M.	"	124°
5:15 P.M.	"	130°

Table 17.4 shows the yields for the sample roasts cooked by this method. Normal conventional shrink in oven roasting would be somewhere between 15 percent and 20 percent, depending on the oven temperature and degree of doneness. Today, with the high cost of top round or even bottom round, the ability to decrease the shrink from a high of 20 percent to somewhere around 10 percent can amount to a savings of $4 to $6 per roast in additional servings per cooked product.

TABLE 17.4   YIELDS FROM SAMPLE ROASTS COOKED IN A WATER BATH

Type of Cut	Gelatin	Initial Weight of Roast	Cooked Weight Direct From Bag	Shrink Loss	%
Top Round	4 oz	8 lb 12 oz	8 lb 1 oz	11 oz	7.8
Top Round	4 oz	9 lb 4 oz	8 lb 7 oz	13 oz	8
Top Round	4 oz	10 lb 2 oz	9 lb 2 oz	16 oz	9.8
Top Round	4 oz	8 lb 6 oz	7 lb 10 oz	12 oz	9
Top Round	4 oz	9 lb 6 oz	8 lb 8 oz	14 oz	9.8

17.4   USDA SAFETY REQUIREMENTS FOR COOKING BEEF AND ROAST BEEF

1.   Cooked beef and roast beef shall be prepared by a cooking procedure that produces a minimum temperature of 145°F (63°C) in all parts of each roast or prepared as provided in items (2) and (3) of this list.

2. Cooked beef may also be prepared by any one of the cooking procedures described in Table 17.5 and in items 3 and 6, and roast beef may also be prepared by any one of the cooking procedures described in the table and in items 4–6 provided that the procedure produces and maintains the minimum temperature required, in all parts of each roast, for at least the stated period.

TABLE 17.5 TIMETABLE FOR AN ALTERNATIVE PROCESSING PROCEDURE FOR COOKED BEEF AND ROAST BEEF

Minimum Internal Temperature		Minimum Processing Time, Minutes
°F	°C	
130	54.4	121
131	55.0	97
132	55.6	77
133	56.1	62
134	56.7	47
135	57.2	37
136	57.8	32
137	58.4	24
138	58.9	19
139	59.5	15
140	60.0	12
141	60.6	10
142	61.1	8
143	61.7	6
144	62.2	5

3. Bag cooking. Each roast to be moist cooked shall be placed in a moisture-impermeable film, either vacuum packaged or with excess air removed, and the bag sealed prior to immersion cooking in a water bath or cooking in an oven.

4. Unbagged cooking (netted or racked roasts). Roasts processed entirely by dry heat must weigh 10 lbs or more before processing and must be dry cooked in an oven maintained at 250°F (121°C) or higher throughout the process.

5. Dry cooking. An oven temperature less than 250°F (121°C) may be used for dry cooking of roasts of any size provided that the relative humidity, as measured in either the chamber or exit vent of the oven in which they are prepared, is greater than 90 percent for at least 25 percent of the total cooking time for the process, but in no case for a shorter period than 1 hour. This relative humidity may be achieved by use of steam injection or by sealed ovens capable of producing and maintaining the required 90 percent relative humidity.

6. A processor who selects any of the alternative procedures specified in paragraphs 2 and 3 must have equipment designed to ensure that beef roasts do not contact each other during processing and shall

have sufficient monitoring equipment to assure that the time (within 1 minute), temperature (within 1°F), and relative humidity (within 5 percent) limits required by this process are being met. The processor shall provide proper recording devices, and make the data from these available to the Food Safety and Quality Service inspection officials upon request. Continuous recording devices with the prescribed accuracies will be acceptable for all products prepared under items (1) and (2).[5]

17.5 SUMMARY

Cooking is a service we provide in the hospitality industry. The guests' perception of a food service facility often hinges on the culinary expertise of the kitchen. Culinary expertise must be built on both technical knowledge and well-developed skills. Food service managers are often weak in this area and rely too heavily on the kitchen staff.

Preparation is a critical link in the system and if it is not evaluated and updated regularly like other areas, it can negate all the management efforts put into purchasing, receiving, and storage. Researchers in the food science area make new discoveries daily which can benefit the food service industry. It is important to remain open to technological advances and to utilize them to serve a more palatable and profitable product to the guests.

CASE STUDY

You are hired as a food service consultant by a steak house chain called Moo. While spending two or three days observing preparation techniques in the production area, you note the following:

- Chef has both left and right sides of the broiler set at 600°F.
- Prime ribs are put in oven at 500°F for one hour then one gallon of water is added to the pan and the temperature is reduced to 325°F.
- Roast pork loins are covered with foil and cooked to an internal temperature of 180°F.
- Average shrink loss on beef roasts amounts to 28 percent.
- Degree of doneness varies from rare to medium well on the prime ribs cooked.
- Thick steaks like filet mignons and strips are very dark on the outside when served.
- Several steaks per night are sent back because they are overcooked.

Your assignment is to analyze these findings and to suggest and demonstrate alternative cooking methods that would convince the preparation staff to alter their procedures.

REFERENCES

1. R.A. Lawrie, *Meat Science,* 3rd ed. (Pergamon Press, 1979), p. 344.

2. E.M. Buck, A.M. Hickey and J. Rosenan, "Low Temperature Air Oven vs. a Water Bath for the Preparation of Rare Beef," *Journal of Food Science,* (1979), vol. 44:1602.

3. L.C. Blankenship, C.E. Davis and G.J. Magner, "Cooking Methods for Elimination of Salmonella Typhimurium Experimental Surface Contaminant for Rare Dry Roasted Beef Roasts," *Journal of Food Science,* (1980), vol 45:270.

4. R.R. Willardson, F.F. Busta, C.E. Allan and L.B. Smith, "Growth and Survival of Clostridium Perfringens During Constantly Rising Temperatures," *Journal of Food Science,* (1978), vol. 43:470.

5. MPI Bulletin, 78–90 8/22/78, United States Department of Agriculture, Food Safety and Quality Service Meat and Poultry Inspection Program, Washington, D.C. 20250.

18 | COSTING AND MENU PRICING

Profitability in a food service operation requires careful cost control and menu pricing. Before menu pricing can be done accurately the exact servable portion cost must be determined. This chapter deals with the determination of those costs by presenting information and illustrations on the cost analysis of fabrication, by explaining the process of yield testing, and by presenting standardized yields for the most frequently used cuts. The chapter concludes with a thorough discussion of the most common methods of menu pricing used in the food service industry today.

18.1   COST ANALYSIS OF FABRICATION

The food service industry once used entire carcasses. It progressed to primal cuts, then to wholesale cuts, and is daily using more and more portion cuts of meat. The reasons for this shift are many. Labor and transportation costs, technological advances in equipment and packaging, and the limited-menu concept are several. Today few operations purchase carcasses and fabricate them on the premises; the exception might be the Kosher operation still purchasing whole forequarters, but since this is a minor segment of the market, it will not be considered here.

Fabrication from the primal level to wholesale cuts and from the wholesale cut level to portion cuts is the extent of fabrication found in most food operations today. Although the trend is clearly away from in-house fabrication toward the purchase of portion cuts, many food operators continue with limited on-premises fabrication. *Why do operators persist in fabrication?* Some claim that it gives them better control over the quality of the steak (they can be sure that the strip loin, for example, is aged for a minimum of two weeks prior to cutting). Second, operators claim they can utilize the production staff more efficiently during off hours by having them fabricate steaks and other cuts. Profit is another frequently mentioned motive. Because of price fluctuation in the fresh meat market, it is possible to generate a savings on some items.

Of course, these arguments can be reversed and used to support the purchase of portion-cut meats. The staff may be fully taxed, leaving no time for fabrication. Or perhaps the controls necessary to do on-premises fabrication make it too difficult to perform, and without the controls additional waste may be generated. So product must be bought in portion-cut form.

Deciding Whether to Fabricate or to Purchase Portion Cuts

There is no one correct choice for all situations. Sometimes fabrication pays and sometimes it does not. The decision to fabricate must be made individually by each management group, taking into consideration both their type of operation and the following factors:
1. Market prices and conditions
2. Labor force
3. Facilities and equipment
4. Use of by-products
5. Controls, storage, training
6. Cost analysis—yield tests

MARKET PRICES AND CONDITIONS—SHORT TERM Because the United States has a fresh meat market where supply and demand fluctuate both nationally and regionally there is a concomitant fluctuation in prices. Since strip steaks are cut from the strip loin, one might expect that a fluctuation up or down in the price of the whole strip loin would affect proportionately the price of the strip steak. Although this is logical, in practice it does not always happen for several reasons. If the purveyor is oversupplied with either of these fresh meat commodities (whole strip loins or strip steaks) the discounting of the item in order to move it and avoid storage costs or loss might be required. Many portion-cut items are now sold in the frozen state, and, because they are prepared from inventory that was purchased earlier, there may be a price fluctuation based on past market cost versus the current market condition. This can be either an advantage or a disadvantage to the purchaser of portion cuts depending on which way the market has moved.

MARKET CYCLES—LONG TERM Market conditions or trends may have another influence on this decision. If the price on strip loins drops dramatically for a short period of time within the market cycle, it might be advantageous to purchase a six-month supply at the low market price thereby locking in an added profit on this item. Interest expense as well as storage and security costs must be calculated before the feasibility of this buying plan can be decided (see bulk buying methods in Chapter 14, Purchasing).

LABOR There are three distinct areas of consideration when labor is evaluated: availability of labor, skill level, and cost of labor.

Availability This term refers to the scheduling of the production crew. Is there any free time during the work day when existing personnel could be utilized to do limited on-premises fabrication? Although many food operators assert that their kitchen staff is tightly scheduled, in reality there is usually a period each day when productivity is very low. This time might possibly be used for fabrication.

Skill level Limited on-premises fabrication does not require a butcher with years of experience. Instead it requires an individual who can be trained to use a steak knife and a portion scale. Experience has demonstrated that high school and college students can accurately and profitably fabricate portion cuts from wholesale cuts such as the strip loin, tenderloin, and top sirloin if properly trained. Initially some product waste is incurred but the skills can be quickly mastered. One benefit is the relatively low cost of labor. Another is that the trainee, whose regular job is often dishwasher, potwasher, or other unskilled worker, develops a feeling of pride in his position because he knows that he is doing something that is extremely important in the operation. This boost for morale has the added benefit of cutting down on turnover of lower level jobs, which concurrently reduces training time and operational difficulties.

Cost of Labor Labor cost is an important consideration. If you are operating a union shop and only the executive chef or an apprentice butcher can do the fabricating, then the cost of the procedure may be prohibitive. If a lower cost employee can be trained to do on-premises fabrication, the cost of labor can be minimized. In some operations it may be feasible to bring in an individual on a part-time basis specifically for the purpose of fabrication, as will be illustrated in some later examples.

FACILITIES AND EQUIPMENT If a food facility is to embark on total fabrication (cutting forequarters, loins, and so on) it definitely needs major equipment additions. A bandsaw would be required, as well as additional walk-in space, rail scales, and other items, which all amount to a sizable capital expenditure. The limited on-premises fabrication under discussion here requires little more than a steak knife, boning knife, portion scale, and, in extreme cases, a handsaw. Quite often these pieces of equipment are already present in a food service facility or can be purchased at a nominal fee should initial purchase or replacement be necessary.

USE OF BY-PRODUCTS Depending on the product to be fabricated, a certain percentage of by-products as well as trim or waste will be produced. In some cases the amount is almost insignificant; but, in other cases, it can be a critical factor in making a decision on fabrication. Consider the example of the primal rib as a product that a limited-menu rib house is investigating for possible fabrication. The primal rib will yield a 109 roast ready rib, with short ribs, stew meat, and ground beef as by-products. If the rib house has no outlet for the by-products, which may amount to 6 lb or more in a 32-lb primal, then the profitability of fabrication is definitely in question. Other products such as the strip loin have very little in the way of by-products and therefore lend themselves to fabrication more frequently.

CONTROLS Management must have the expertise required to control the fabrication process to ensure uniform portions and prevent pilferage.

Standardized yields must be determined, and management must check the results of fabrication to be sure the individual cutting is accurate and honest. If the person fabricating decides to steal two steaks he may undercut (light portion) 10 or 12 steaks to make up for it. If portion weights are not checked, this may not show up until the guests complain about small portions. Once the steaks are cut accurately, the same controls that apply to items purchased in portion-controlled form must be exercised (they should be issued by count and checked against dining room checks at end of the shift).

STORAGE CONTROLS AND TIME PERIOD The turnover rate of fabricated products must be checked by management since packaging available for in-house fabrication is limited to hand wrapping with plastic wrap. Refrigerated shelf life for individual portions wrapped in this manner is only 3 to 5 days. This should allow ample storage time if proper inventory rotation is practiced, however. The motto for inventory rotation is **FIFO**—first in, first out—and the object is to use any product before it reaches the end of its good-quality keeping time.

Wholesale cuts can be kept much longer (1–2 weeks), and during slow periods fabrication may be performed daily or as needed from the wholesale cut.

Modern packaging techniques and materials (Bivac, Multivac, Cryovac) available on products portion-cut by purveyors give extended shelf life to individual portions. If forecasting and inventory turnover cannot be controlled adequately by management, then packaging material may also be a consideration in management's decision to cut in-house or to purchase portion-cut items.

TRAINING The fabrication procedure to be described in this chapter is basically simple and can be performed by anyone with normal motor skills, but the training must be done by someone who has the technical knowledge to fabricate properly. If this knowledge is not available within the operation, an expert can be brought in for a training program or an employee can be sent to a seminar or course on meat cutting. In any case, after the individual has been instructed in fabricating skills and has applied this information on the job, there should be a follow-up training session to assure adequate knowledge and competence.

COST ANALYSIS—YIELD TESTS Once the factors of market, labor, equipment, usage, and controls have been considered, the operator must put the fabrication question to the final test. Yield tests must be performed on the various primal and wholesale cuts to determine profitability.

The following section presents examples of primals fabricated to wholesale cuts and wholesale cuts fabricated to portion cuts. They show the process involved in performing a yield test and tie together the decision factors just listed. Table 18.1 is a sample yield test form.

TABLE 18.1 A YIELD TEST FORM

Wholesale Cut_____   Date _____

Weight _____   Cost/lb_____Total_____

Grade_____   Purveyor _____

Breakdown	Weight lb oz	Value per lb	Total Value	Purchase price
Usable trim				↓
Waste				Less value of usable trim
Fat				
Bone				↓
Primary product				Equals value of primary product before labor

Labor to fabricate	Time	Rate/Hour	Total	
				↓
				Plus labor
				↓
				Equals total cost of fabricating product
				↓
				Portion Cost

Total cost

Divided by
number of portions

Test performed by

Comments

18.2 YIELD TESTING

A Yield Test Formula

Weight of cut as purchased × price/lb = purchase price,
  less value of usable trim,
  plus cost of labor to fabricate,
  equals cost of fabricated product, $A.

This cost is then compared to the cost of buying the product prefabricated, calculated as follows:

Weight of fabricated product × Market price/lb = Cost to purchase prefabricated product, $B.

If A is less than B, a profit on fabrication is made. To illustrate with actual figures, fabrication of a 109 rib from a 103 will be compared to purchasing a 109 prefabricated.

Choice NAMP #103 primal rib 35 lb @ $1.30 = $45.50 purchase price.

Breakdown Yields		Market Price	Value
4.5 lb short ribs	@	$1.25	$ 5.63
1.5 lb ground beef	@	1.30	1.95
2.0 lb stew meat		1.60	3.20
7.0 lb fat and bone		No value	
		Total value of by-products	$10.78

To compare fabricating a 109 (20 lb) to direct purchasing a 109:

$45.50	cost of primal
−10.78	less value of by-products (used elsewhere)
34.72	
+ 1.50	labor (value added) ¼ hour @ $6/hour
$36.22	cost of 109 fabricated for 103

Market value for 109 is $1.95 × 20 lb = $39.00. Savings = $2.78 if by-products can be utilized. If not, loss = $8.00.

The NAMP *Meat Buyer's Guide* gives details on the breakdown of the 103 primal rib to the 109 roast ready rib, with the by-products obtained from fabrication. Depending on how the by-products are used, different values would be assigned to them. For example, if short ribs could not be utilized, the meat obtained from the short ribs would be put into ground beef, thus affecting the yield and value of the by-products. If the operation has a need for the by-products and they can be effectively used on the menu (or for employees' meals), then a potential savings of

$2.78 per rib is possible under these market and labor costs. A savings of $2.78 may not seem very significant, but remember that this $2.78 is in fact increased profit, since labor and other costs have already been taken into consideration.

Note that the labor in this example was added into the cost of the fabricated 109. If, in fact, regular kitchen staff was utilized during slow periods to fabricate, this cost would not really be an addition to the payroll and, therefore, the process would generate a savings of $4.28 instead of $2.78. The labor cost is included to show that even if outside help is brought in, fabrication can be profitable. (Some operations hire a butcher from a local supermarket one day per week to do fabrication.)

Fabricating primal rib to 109 requires the most extensive amount of fabrication that the average food service operation would consider. Much simpler fabrication with limited by-products is considered in the following example of the strip loin.

Choice NAMP #180 strip loin 12 lb @ $2.80/lb = $33.60.

Breakdown Yields	Market Price	Value
1.5 lb waste	No Value	
1.5 ground beef	$1.30	$1.95
9 lb (12 12-oz steaks ¼ in. fat cover, 1½ in. tail)		

To compare fabricated steaks to portion cuts:

$33.60	cost of strip loin
− 1.95	by-product
31.65	
+ 1.50	labor ¼ hour @ $6.00
$33.15	cost of steaks fabricated from strip loin

$33.15   ÷12 steaks = $2.76/steak

If Choice portion-cut steaks cut end to end are $0.27/oz or $3.24/steak on the market, fabrication saves $0.48 steak or $6.00 strip loin.

Determining the yield of portion-cut steaks from a whole strip loin

To illustrate the procedure in performing a yield test the following steps applicable to fabrication of the strip loin are listed:
1. Weigh whole strip loin.
2. Cut tail off at length desired for steak.
3. Trim fat cover to acceptable tolerance for steaks.
4. Start cutting from rib end, not loin end (the connective tissue at the loin end makes it less desirable).
5. Look at the width of eye muscle and estimate the thickness needed to cut proper portion size. It is better to cut the first steak slightly overweight to ensure usable portion.

6. Use a portion scale and adjust thickness to produce uniform portion weight as dimensions of strip eye change.

7. One-half-oz tolerance is generally acceptable on portion sizes of 8 oz and up.

For a strip loin, the Choice 180 strip loin is cut end to end after trimming to the specifications noted. It yields 12 12-oz steaks costing $2.76 per steak. The portion-cut prices are $0.27 per ounce or $3.24 per steak, giving a profit of $0.48 per steak or $6.00 per strip loin. Once again, the figure should be considered profit. It is interesting to note that if the strip steak sells on the menu for $10.00 à la carte and the national averages for profit percentage on foods are applied (5%), the profit on that strip loin steak is approximately $0.50. If fabrication is performed and the savings is $0.48 per steak, the profit is almost doubled (to $0.98) by this process.

Accurate comparison requires detailed specifications

When doing the yield test, it is imperative to compare the same specification steak (grade, trim, tail). If only center cut steaks were to be purchased, the price per ounce would be more than shown in the example. In this case, the wholesale strip would yield more in the way of by-products from the loin end, which contains the connective tissue. At least two steaks would be eliminated and have to be used for beef tips or some other product. This may affect the outcome if the operation has no use for the by-products. Also consider what happens if two or three steaks are improperly cut. If only nine steaks of the proper weight and trim are obtained, the cost per steak is $33.15 ÷ 9 or $3.68 (assuming mis-cuts cannot be utilized. Using steaks that are over or under the proper weight range leads to customer dissatisfaction. Steaks from a purveyor would be rejected if they did not meet the specified weight range; they must also be rejected when in-house fabrication fails to meet specifications. If they are not, there is no accurate way to compare portion cuts with cuts fabricated in-house.

Experience shows that the most profitable beef cuts to fabricate on the premises are the strip loin, tenderloin, and the top sirloin. Lamb and veal legs, turkey breasts, and broilers and fryers are other potentially profitable items to fabricate. The only way for management to make a final decision on whether fabrication will be profitable is to do an actual yield test. After several tests have been completed using the same grade, weight range, and specifications, standardized yields can be developed, showing percentage breakdowns of the primary products versus the by-products of fabrication. These percentages then can be applied to current and future market prices to determine profitability.

Average yields of major primal and wholesale cuts

Listed in Table 18.2 are average percentage breakdowns for the major wholesale cuts used by the food service industry. These figures were compiled with the help of three major companies: Cross Brothers, Philadelphia, a slaughterer; Otto Salmon and Sons, a major breaker in New York; and Colonial Beef, a portion-control company in Philadelphia. The tonnage of product tested by these companies makes the statistics much more accurate than small quantity tests. Please keep in mind that actual yields may vary from these because of different trim factors, weight ranges, and so on. These statistics should be used only as a guide. By applying current market prices to the yields listed and by considering the other three factors (labor, facilities, and use of by-products), management can make a sound decision on in-house fabrication.

TABLE 18.2 MAJOR BEEF CUTS—PRIMALS TO WHOLESALE CUTS AND WHOLESALE TO PORTION CUTS

Primal rib (103) to sub primals, 103–109 roast-ready rib From #103

Primal Rib (weight range 28–33 and 33–38)

Breakdown Yields			30 lb	32 lb	34 lb	36 lb	38 lb
To #109	rib, roast ready	57%	17.1 lb	18.24	19.38	20.52	21.66
	short ribs	6.45%	1.9	2.06	2.19	2.32	2.45
	ground beef	15.43%	4.6	4.94	5.25	5.55	5.86
	stew meat	6.67%	2.0	2.13	2.27	2.40	2.53
	waste	14.65%	4.4	4.7	4.98	5.27	5.57

Primal rib 103 to 107 rib, oven-prepared

From #103 primal rib (weight range 28–33 and 33–38)

Breakdown Yields			30 lb	32 lb	34 lb	36 lb	38 lb
To #107	rib, oven-prepared	70%	21 lb	22.4	23.8	25.2	26.6
	short ribs	6.5%	1.95	2.08	2.21	2.34	2.47
	fat and bone	23.5%	7.03	7.52	7.99	8.46	8.93

Primal rib 103 to 112

From #103 primal rib (weight range 28–33 and 33–38)

Breakdown Yields			30 lb	32 lb	34 lb	36 lb	38 lb
To #112	rib eye roll	25%	7.5 lb	8	8.5	9	9.5

From 172 full loin to wholesale cuts

From #172 full loin (weight range 48–52, 53–57, and 58–62)

Breakdown Yields			50 lb	55 lb	60 lb
To #180	strip loin	18.60%	9.3 lb	10.23	11.15
#189	tenderloin	9.06%	4.53	4.98	5.44
#184	top sirloin butt	17.05%	8.52	9.37	10.23
#185	bottom sirloin butt	10.07%	5.03	5.54	6.04

From #175 bone-in strip loin to #180

From #175 strip loin (weight range 13–16, 16–19 lb)

Breakdown Yields			15 lb	16 lb	17 lb	18 lb
To #180	strip loin	66%	9.9 lb	10.56	11.22	11.88
	lean trim	7%	1.05	1.12	1.19	1.26
	fat and trim	2%	.30	.32	.34	.36
	bones	10%	1.50	1.60	1.70	1.80
	fat	11%	1.65	1.76	1.87	1.98

From #189 tenderloin to #189A.

From #189 full tenderloin (weight range 5–7 lb)

Breakdown Yields		5 lb	6 lb	7 lb
To #189A full tenderloin, defatted, side muscle on	70%	3.5 lb	4.2	4.9

From #189 full tenderloin (weight range 5–7 lb)

Breakdown Yields		5 lb	6 lb	7 lb
To #190 full tenderloin special	60%	3 lb	3.6	4.2

From #180 strip loin to strip steaks.

Center cut 60%
End to end 73%

From #180 strip loin (weight range 10–12 lb)

Yields	Center cut 60%			End to end 73%		
	10 lb	11 lb	12 lb	10 lb	11 lb	12 lb
	6 lb	6.6	7.2 lb	7.3	8.03	8.76
To 8-oz. steaks	12	13	14	14	16	17
# 10-oz steaks	9	10	11	11	12	14
# 12-oz steaks	8	8	9	9	10	11

From top sirloin butt #184 to steaks

Center cut steak	44% yield	End-to-end steak	60% yield
	25% trim		12% trim
	31% fat		28% fat

From top sirloin butt #184 (weight range 10–12 lb)

Yields	Center cut 44%				End to end 60%		
	10 lb	11 lb	12 lb		10 lb	11 lb	12 lb
	4.4 lb	4.84	5.28		6 lb	6.6	7.2
To 6-oz steaks	11	12	14		16	17	19
# 8-oz steaks	8	9	10		12	13	14
# 10-oz steaks	7	7	8		9	10	11
Trim and deckle (25%)	2.5 lb	2.75	3	(12%)	1.2 lb	1.32	1.44
Fat (31%)	3.1 lb	3.4	3.72	(28%)	2.8 lb	3.08	3.36

From #173 short loin to 14-oz club, T-bone and Porterhouse steaks

Yields (14-oz)		Short loin weight range	
		21–25 lb	25–28 lb
club steaks	25%	6	8
T-bone	11%	3	4
porterhouse	30%	8	9

From #158 primal round to #168

From #158 primal round (weight range 70–95 lb)

Breakdown Yields			70 lb	75 lb	80 lb	85 lb	90 lb	95 lb
To #168	top round	24.63%	17.24 lb	18.47	19.70	20.94	22.17	23.4
	gooseneck	31.89%	22.32	23.92	25.51	27.1	28.7	30.3
	knuckle	16.33%	11.43	12.25	13.06	13.88	14.7	15.51
	trim and waste	18.24%	12.77	13.68	14.59	15.50	16.42	17.33
	shank meat	5.85%	4.09	4.39	4.68	4.97	5.27	5.56

Veal leg, bone-in to veal cutlets

From #233 veal leg-bone-in

Breakdown Yields		14 lb	16 lb	18 lb	20 lb	22 lb
cutlets	51%	7.14 lb	8.16	9.18	10.2	11.22
trim	26%	3.64	4.16	4.68	5.2	5.72
waste	23%	3.22	3.68	4.14	4.6	5.06

TABLE 18.3 MAJOR WHOLESALE CUTS OF LAMB

Hotel rack of lamb to rib chops

From hotel rack of lamb (weight range 5–7 lb)

Yields		5 lb	6 lb	7 lb
To 4 oz chops	57%	2.85 lb	3.42	3.99
# chops		11	13	15
6 oz chops	59%	2.95 lb	3.54	4.13
# chops		7	9	11
trim	5%	0.25 lb	0.3	0.35
waste	38%	1.9 lb	2.28	2.66

Lamb loin to loin chops

From lamb loin (weight range 6–10 lb)

Breakdown Yields		6 lb	7 lb	8 lb	9 lb	10 lb
To 4-oz chops	50%	3 lb	3.5	4	4.5	5
# chops		12	14	16	18	20
6-oz chops	54%	3.24 lb	3.78	4.32	4.86	5.4
# chops		8	10	11	12	14
trim	5%	0.3 lb	0.35	0.4	0.45	0.5
waste						
on 6 oz	40%	2.4 lb	2.8	3.2	3.6	4
on 4 oz	45%	2.7 lb	3.15	3.6	4.05	4.5

Leg of lamb, bone-in to boned and tied leg.

From #233 lamb leg-bone-in

Breakdown	Yields		7 lb	8 lb	9 lb	10 lb	11 lb
To #234A	leg boned and tied	72%	5.04 lb	5.76	6.5	7.2	7.9
	trim	2%	1.4	1.6	1.8	2	2.2
	waste	26%	1.8	2.1	2.3	2.6	2.8

18.3 MENU PRICING SYSTEMS

Once management has made a decision concerning fabrication on the premises versus purchasing portion cuts, the next logical step is pricing the item on the menu. Several methods of menu pricing are practiced today. Each method has its advantages and disadvantages and should be considered carefully before being used. The following examples are only a sampling of the possible methods for pricing entrees on the menu.

Straight Percentage or Factor Method

This model for menu pricing is used often and is based on a desired food cost percentage. The formula is as follows:

$$\frac{\text{Raw food cost}}{\text{Desired food cost percentage}} = \text{Menu sales price}$$

or

Raw food cost $\times$ Pricing factor = Menu sales price, where the pricing factor is 1/Food Cost percentage.

For example, if the operation desires a 36% food cost percentage, a 12-oz strip steak (12-oz raw weight) which costs $2.85/portion, would be priced about $8.00.

$$\frac{\$2.85 \text{ (Raw food cost)}}{36\% \text{ (Desired food cost \%)}} = \$7.92 \text{ (Menu selling price)}$$

or

$$\frac{1}{36\% \text{ (f.c.\%)}} = 2.77.$$

$2.85 (Raw food cost) $\times$ 2.77 (Pricing factor) = $7.90

This method is easy to use, but fails to consider whether the remaining 64% of income will cover all operational costs and still yield a profit at the end of an accounting period. The method may not be suitable for certain items. For example, using the same 36% food cost, a chicken entree that costs $0.80/portion would be priced $2.25 on the menu, while a lobster entree that costs $6.00/portion would be priced about $16.75 on the menu.

Major Cost Method (Prime Cost)

Another method considers the major costs in running a food service operation—food and labor—in calculating the selling price. The formula is in four steps:

Step 1. $\dfrac{\text{Labor cost for month or year}}{\text{\# of covers/month or year}}$ = Labor cost/Meal served.

Step 2. Cost of food + Labor cost/Meal = Major costs.

Step 3. Determine desired major cost %.

Step 4. $\dfrac{\text{Major costs}}{\text{Major cost \%}}$ = Selling price.

For example, if a restaurant served 48,000 covers per year, sales were $480,000/year, and labor costs were $150,000/year (32% labor cost derived from income statement), the same strip steak ($2.85/portion) would be priced at about $8.75, calculated as follows:

1. $\dfrac{\$150,000 \text{ (labor cost/year)}}{48,000 \text{ (\# covers/year)}}$ = $3.11 (Labor cost/Meal served).

2. $2.85 (Cost of food) + $3.11 (Labor cost/Meal) = $5.96 (Major costs).

3. 36% (Desired food cost %) + 32% (labor cost) = 68% (Desired major cost %)

4. $\dfrac{\$5.96 \text{ (Major costs)}}{68\% \text{ (Major cost \%)}}$ = $8.76.

Although this method includes labor costs, it still does not consider the overhead costs per meal (administrative wages, rent, utilities) or guarantee a certain profit percentage. By determining the overhead cost per meal, a third method can be developed as follows.

Gross Cost Method—Profit Built In

Step 1. Raw food cost + Labor cost/Meal + Overhead/meal = Gross Costs.

Step 2. Food cost % + Labor cost % + Overhead cost % = Gross cost %.

Step 3. $\dfrac{\text{Gross costs}}{\text{Gross cost \%}}$ = Selling price.

Using the previous example, if overhead costs were $120,000 (25% derived from income statement), then overhead cost per meal would be $2.50 ($\frac{\$120,000}{48,000}$ = $2.50). Using the gross cost formula, the steak would be priced at $9.00. Not only does this method include the overhead costs per meal, but it also builds a 7% profit into the menu price.

1. $2.85 (Raw food cost) + $3.11 (Labor cost/meal) + $2.50 (Overhead/Meal) = $8.46 (Gross costs).

2. 36% (Food cost) + 32% (Labor cost) + 25% (Overhead cost) = 93% (Gross cost).

3. $\dfrac{\$8.46 \text{ (gross costs)}}{93\% \text{ (gross cost \%)}}$ = $9.09 (Selling price).

Note: The labor costs calculated in this problem represent an average. Not all items require the same preparation time nor need to be assigned the same cost: strip steak only requires broiling but chicken Cordon Bleu requires fabrication time, stuffing, browning, and baking. Therefore, maybe the steak should be assigned a $2.25 labor cost and the Cordon Bleu a $4.00 labor cost.

Actual Cost Method

Still another method, referred to as the actual cost method, considers the raw food cost in dollars, labor cost in dollars, variable cost in percentage of sales, and desired profit in percentage of sales to determine the menu selling price. The following example for pricing a strip loin steak outlines this procedure.

Step 1. Determine cost
a. Given food cost   = $2.85
b. Labor cost   = $3.11
c. Total food and labor cost = $5.96

To determine the labor cost use a time and motion study or determine the percentage of labor cost from the profit and loss (P&L) statement .

Step 2. Determine variable cost, fixed cost, and profit.
a. Variable cost   =   9% of sales
b. Fixed cost   = 16% of sales
c. Estimated profit = 10% of sales
                35%

The variable cost percent (vc%) = vc/s, or *variable costs* (utilities, china, glass, silver, etc.) divided by gross sales. The fixed cost percent is determined by dividing the fixed costs (rent, insurance, etc.) by sales. fc/s = fixed cost percent.

Step 3. Determine selling price
a. Selling price = X or 100%
b. Fixed cost + Variable cost + Profit = 35%
c. Food + Labor = $5.96
d. Food + Labor = 100% – 35% or 65%
e. 65% = $5.96

f. $X = \dfrac{\$5.96}{65\%} = \$9.17$

g. $X = \$9.15$ menu price

These calculations show that at a selling price of $9.15, a profit of $0.90 will be made on this item. This method can also be used to help an operator know what new products may be successfully marketed on the menu. If customer resistance is evident on items priced above this $9.15 level, then management must look for menu items with a food cost of less than $2.85 in order to have both volume and the desired profit margin.

Volume/Risk/Profit Method

The last method, an adaptation of the Texas Restaurant Association's system by Jack E. Miller in his book *Menu Pricing*,[1] is based on volume, popularity, and profit mark-up. It categorizes menu items by volume, risk and cost, as follows:

Volume/Cost Categories

1. High volume + high cost = popular item but not profitable.
2. High volume + low cost = popular item and profitable.
3. Low volume + high cost = unpopular item and not profitable.
4. Low volume + low cost = unpopular item and profitable.

Menu Categories	Profit Mark-up Range, %
Appetizers	20–50
Salads	10–40
Entrees	10–25
Vegetables	25–50
Beverages and breads	10–20
Desserts	15–35

According to Miller, there are three basic steps to follow. First, the operator must determine the menu selling price (see the following example), then through sales analysis determine which of the four categories it falls into, and then finally select the appropriate profit mark-up based on the item's popularity and profitability.

Step 1. Determine percentages of overhead, labor, and desired profit.

Using previous examples, overhead is 25%, labor is 32%, and profit is 7%, which totals 64%. The remaining 36% is the food cost (100 – 64 = 36).

Step 2. Determine raw food cost.

Using the previous example, the strip steak costs $2.85/12-oz. portion.

Step 3. Divide raw food cost by the food cost % to get menu selling price.

$$\frac{\$2.85 \ (\text{Raw food cost})}{36\% \ (\text{Food cost \%})} = \$7.92. \quad (\text{Base price is } \$7.92.)$$

Once the food operator determines the base selling price, the menu item must be assigned to one of the four volume/cost categories. For this example, assume the strip steak falls into category (1)—it is a high-volume, high-cost, popular item and not profitable. Using the profit mark-up chart, the operator would then add an extra 15% (due to medium risk) to the menu selling price and the steak would now be priced between $9.00 and $9.25 (15% of $7.92 = $1.19, $1.19 + $7.92 = $9.11). Menu items fitting into categories (1) and (4) would receive a medium profit mark-up, and those items fitting into categories (2) and (3) would receive a low profit mark-up and high profit mark-up, respectively.

18.4 SUMMARY

The market is the key to menu planning. Ultimately, the customer dictates what prices can be charged on a menu and what items will sell. The food service operator must keep the potential market in mind when selecting menu items, making decisions on fabrication, and pricing the menu. This chapter provides the technical tools and the managerial approach to the complex topics of on-the-premises fabrication and menu pricing. The information provided should be helpful to both management and staff for evaluating product yields and profitability.

References

1. Jack E. Miller, *Menu Pricing* (Boston: CBI Publishing Co. Inc., 1976) pp. 14–15.

19 | THE FUTURE

At this writing, the beef cycle is in its rebuilding phase. The normal cattle cycle takes between 6 and 12 years to complete. In 1968 there were 109 million cattle in the United States. Demand was good as income levels rose, signaling producers to increase the herds. By 1973 the number of cattle had increased to 133 million. Two things then happened that influenced beef production: consumers boycotted beef because of higher prices, and short-term wage and price controls were imposed. As a consequence farmers stopped breeding cattle and began liquidating stock. During the liquidation phase, 1973–78, there was an abundance of lean meat on the market (created by the slaughter of breeding stock) and this kept prices suppressed. But the same years found the rapid expansion of the fast food industry, with many chains doubling their number of units. By 1979 the number of cattle had been reduced to approximately 110 million. With the great demand that had built up during the liquidation period, prices had to rise when the supply was diminished. The price rise in turn signaled the producer to start the rebuilding cycle.

Increasing the beef herds takes a minimum of four years. First, the cow must conceive. The gestation period is 10 months, then the growth period is about 16 months before that offspring can be bred. (If that first calf is slaughtered, the herd never increases.) Then another 10-month gestation period followed by 16–18 months growth until slaughter. This makes a total of 54 months before significant increase in herd size can be made. Once the numbers are high enough to influence the supply/demand price balance, it is likely that a new liquidation phase will begin.

With high beef prices, consumers look to poultry and pork to supply more of their meat needs at home. Pork and poultry production can be increased much more quickly than beef production (pork in 6–8 months, poultry in 10–12 weeks), and these two meats will undoubtedly take up much of the slack caused by low beef supplies.

Domestic lamb and formula-fed veal are also in short supply. They are rapidly becoming luxury items consumed more at restaurants than in the home. Shellfish prices continue to rise, but fin fish have shown some stability in prices with increased catches attributed to the 200-mile limit.

What the future holds in the way of supplies is difficult to say without a crystal ball. But the experimental work going on today in food science laboratories may point to some of the changes we will see in future supplies. The following list is by no means all inclusive, but it is offered to give the reader some idea of what is being done in different areas:

FLAKE FORMED AND RESTRUCTURED MEATS A lower grade product is put through a device that slices it into flakes rather than grinding it. These flakes can then be bound together under pressure to form steaks, chops, or roasts. By using various size dies and different amounts of pressure, the texture and appearance of many products can be simulated. This system is already in production in some plants and it should see increased use as prices continue to rise.

NEW PORTION CUTS Advances in technology such as the packaging systems of boxed beef, blade tenderizing, and better freezing systems have allowed portion-control companies to develop new products that are alternatives to the high-priced strip steaks and filet mignon. Colonial Beef, for example, in Philadelphia, has developed a sirloin butt strip and a filet of sirloin from the bottom sirloin butt. Other sections of the carcass are now being tested to develop additional products.

The rise of mechanical equipment such as hydraulic presses and mechanical steak cutters has made portion cuts much more uniform and economical. A skilled butcher can cut approximately 550 steaks per hour compared to a mechanical steak cutter, which can accurately portion cut over 220 steaks per minute to within $\frac{1}{8}$-ounce tolerance.

AQUACULTURE Recently much research has been done on the culturing of shrimp and lobsters. Major breakthroughs have been made in developing pellet feeds and in reducing significantly the amount of time lobster and shrimp take to reach market weight. Advanced work in this area is needed, but the hope is for a constant supply and stable price for these very high demand items. Under natural conditions it takes six years for a lobster to reach market weight of one pound. A female lobster will produce 30,000 eggs, but less than one percent will survive to reach adult size. (The eggs are eaten by fish and birds, washed ashore, and so on.) A one-pound lobster can be produced by aquaculture in 2 to 2½ years, but it costs about 2½ times the current market price for lobster. Therefore it is not economically feasible at this time. It is possible to increase the natural survival rate by raising the eggs through the stage of shell formation and then releasing the tiny but fully formed lobsters into the sea. This should stabilize future catches until better feed efficiencies can be produced, which may make feasible raising the lobster to market size in captivity.

The technology for raising fin fish under aquacultural conditions has already been developed and trout and catfish are being raised on fish farms. Oysters and some clams are also being produced by aquaculture, and, as natural stocks become depleted or polluted, there will undoubtedly be more reliance on aquaculture.

MINCED FISH Only 240 species of fish are marketed today. Yet there are thousands of species in the world's lakes, rivers, and oceans. Many types of fish are not harvested because they have been given undesirable names or because they contain too many bones. Often these fish are classified as trash fish because people throw them out rather than eat them.

Dr. Robert Baker at Cornell University working under a grant from the Office of Sea Grant of the National Oceanic and Atmospheric Admiristration (NOAA) has developed a method of mechanically deboning underutilized species. He used the Beehive deboner to produce totally boneless and skinless minced fish, a product similar in appearance to finely ground veal or poultry. It was test marketed at retail in 1-pound units at the same price as hamburger. Dr. Baker has also developed a number of convenience items from the minced fish: fish chowder, fish crispies (resembling scallops), Sloppy Jonahs, fish hot dogs, and other items.

The uses for this product in the food service industry are limitless and the cost is very low. This product is used for appetizers, soups and entrees. Seafood crepes, Newburg, creole, *au gratin*, stuffings, and bouillabaisse are just a few of the items in which this product could be used.

MINCED POULTRY   Dr. Baker also utilized the deboner on spent laying hens to produce minced chicken. The minced chicken was incorporated into convenience products such as chili, lasagna, spaghetti and meatballs, pizza, tacos, enchiladas, and chicken hot dogs. Minced poultry can be utilized in almost all dishes where ground beef is normally used. And its price is less than half the price of ground beef.

CONVENIENCE ITEMS   Advances in freezing and packaging have paved the way for quality convenience food entrees. The entrees have most of the preparation labor already performed, and they require minimum effort and time to reconstitute. Products such as Beef Wellington, Veal Cordon Bleu, Chicken Kiev, Quiche Lorraine are being produced by several reputable suppliers. These products simplify purchasing, receiving, storage, and pricing as well as preparation. Fully cooked items such as beef roasts, turkey breasts, and so on, which have had not only the labor and energy costs controlled but also the shrink losses reduced provide the food service operation with an efficient ready-to-eat product. As development of these products continues, it will certainly have an effect on the staffing and equipment needs of kitchens designed for the future.

SUMMARY

As other nations gain in wealth, the worldwide demand for meat will grow. The U.S. will find itself in a very competitive marketplace and may find little available meat to import. This situation will put even more pressure on the domestic market and will necessitate better utilization of our supplies.

This surely will stimulate additional technological changes, which will produce even more new products and systems than those recently developed. The food service operator should remain open to new developments, testing new products and keeping abreast of the technical changes. It is the food service manager's job to apply this technology to all management areas discussed in this book—purchasing, receiving, storage, and preparation of meat, poultry, and seafood.

SUGGESTIONS FOR FURTHER READING

Anderson, B.B., L.L. Robinson, and J.E. Hodgkins. "The Banco Test: A Rapid Method for Fat in Meat and Edible Meat Products." *J. Assoc. Off. Agr. Chem.,* 45:13, 1962.

Anderson, M.E., R.T. Marshall, W.C. Stringer, and H.D. Naumann. "Efficacies of Three Sanitizers Under Six Conditions of Application to Surfaces of Beef." *J. Food Sci.,* 42:326, 1977.

Ashby, B.H., and G.M. James. "Effects of Freezing and Packaging Methods on Shrinkage." *Journal of Food Science,* vol. 39, 1974.

Ashrae. *Refrigeration Applications.* New York: American Society of Heating, Refrigeration and Air-Conditioning Engineers, 1974.

Bardach, John E., J.H. Ryther, and W.O. McLarney. *Aquaculture: The Farming and Husbandry of Freshwater and Marine Organisms.* New York: Wiley, 1972.

Beef and Veal—The Good Cook/Techniques and Recipes. Alexandria, Virginia: Time-Life Books, 1979.

Behnke, James R. "Freezing from Symposium: Innovations in Processing of Refrigerated and Frozen Foods." *Food Technology,* December, 1976.

Bengtsson, et al. "Cooking of Beef by Oven Roasting: A Study of Heat and Mass Transfer." *J. Food Sci.,* vol. 42, 1977.

Berry, B.W., G.C. Smith, and Z.L. Carpenter. "Beef Carcass Maturity Indicators and Palatability Attributes." *J. Anim. Sci,* 38:507, 1974.

Bidner, T.D. "Update: A Comparison of Forage-Finished and Grain-Finished Beef." *Proceedings 28th Annual Reciprocal Meat Conference of The American Meat Science Association.* Chicago: National Live Stock and Meat Board, 1975.

Birmingham, E., H.B. Naumann, and H.B. Hedrick. "Relation Between Muscle Firmness and Fresh Meat Stability." *Food Technol.,* 20:1222, 1966.

Borton, R.J., N.B. Webb, and L.J. Bratzler. "A Research Note. Influence of Odor Sources on the Odor and Flavor of Beef." *J. Food Sci.,* 39:424, 1974.

Bouton, Harris, and Shorthose. "Factors Influencing Cooling Losses from Meat." *J. Food Sci.,* Sept./Oct., 1976.

Bunch, W.L., M.E. Matthews, and E.H. Marth. "Hospital Chill Food-Service Systems: Acceptability and Microbiological Characteristics of Beef-Soy Loaves When Processed According to System Procedures." *J. Food Sci.,* 41:1273, 1976.

Campion, D.R., J.D. Crouse, and M.E. Dikeman. "A Research Note. The Armour Tenderometer as a Predictor of Cooked Meat Tenderness." *J. Food Sci.,* 40:886, 1975.

Carpenter, Z.L., R.G. Kauffman, R.W. Bray, and K.G. Weckel. "Factors Influencing Quality in Pork. B. Commercially Cured Bacon." *J. Anim. Sci.,* 28(5):578, 1963.

Carpenter Z.L., G.C. Smith, and W.D. Farr. "Reflections on the Beef Grade Changes." *Proceedings of the Meat Industry Research Conference,* March, 1977.

Clark, D.S., and T. Burki. "Oxygen Requirements of Strains of Pseudomonas and Achromobacter." *Can. J. Microbiol.,* 18:321–326, 1972.

Copson, David A. *Microwave Heating,* 2nd ed. Westport, Connecticut: Avi Publishing Co., 1975.

Cross, H.R., E.C. Green, M.S. Stanfield, and W.J. Franks, Jr. "Effect of Quality Grade and Cut Formulation on the Palatability of Ground Beef Patties." *J. Food Sci.,* 41:9, 1976.

Cross, H.R., M.S. Stanfield, E.C. Green, J.M. Heinemeyer, and A.B. Hollick. "Effect of Fat and Textured Soy Protein Content on Consumer Acceptance of Ground Beef." *J. Food Sci.,* 40:1331, 1975.

Davey, C.L., and K.V. Gilbert. "Cooking Shortening and the Toughening of Beef." *J. Food Technol.,* 10:333, 1975.

Davey, C.L., and K.V. Gilbert. "The Tenderness of Cooked and Raw Meat from Young and Old Beef Animals." *J. Sci. Food Agric.,* 26:953, 1975.

Decareau, R.V. "Microwave Energy in Food Processing Applications." *CRC Crit. Rev. Food Technol.,* 1:199, 1970.

DeMan, J.M., and P. Melnychyn. *Symposium: Phosphates in Food Processing.* Westport, Connecticut: Avi Publishing Co., 1971.

Desrosier, Norman W., and Donald K. Tressler. *Fundamentals of Food Freezing.* Westport, Connecticut: Avi Publishing Co., 1977.

Durocher, Joseph F., and Raymond J. Goodman, Jr. *The Essentials of Tableside Cookery.* Ithaca, New York: Cornell University School of Hotel Administration.

Dutson, T.R., G.C. Smith, R.L. Hostetler, and Z.L. Carpenter. "Post-Mortem Carcass Temperature and Beef Tenderness." *J. Anim. Sci.,* 41:289 (Abstr.), 1975.

Ernst, Lyle J. "Changing Demands for Meat." *Proceedings of the Meat Industry Research Conference*, March, 1979.

Fergusson, Jeremy L. "Existing Products for Microwave Ovens." *Food Engineering,* September, 1976, pp. EF23–24.

Forrest, J.C., E.D. Aberle, H.B. Hedrick, M.D. Judge, and R.A. Merkel. *Principles of Meat Science.* San Francisco: W.H. Freeman, 1975.

Harrell, Bidner, and Icazu. "Effect of Altered Muscle pH on Beef Tenderness." *J. Anim. Sci.,* 46(6):1592, 1978.

Ho, Chi-Tang. "Meat Flavor: Past, Present and Future," *Proceedings of the Meat Industry Research Conference,* March, 1978.

Hornstein, I., and P.F. Crowe. "Flavor Studies on Beef and Pork." *J. Agric. Food Chem.,* 8:494, 1960.

Hughes, John T. *Grow Your Own Lobsters Commercially.* Commonwealth of Massachusetts, Publication #10265-5-1000-2-78-CR.

Jones, K.B., J.M. Harries, J. Robertson, and J.M. Akers. "Studies in Beef Quality, IV, A Comparison of the Eating Quality of Meat from Bulls and Steers," *J. Sci. Food Agric.,* 15:790, 1964.

Juillerat, M.E., and R.F. Kelly. "Quality Traits Associated with Consumer Preference for Beef." *J. Food Sci.,* 36:770, 1971.

Kahn, Lentz. "Influence of Ante-Mortem Glycolysis and Dephosphorulation of High Energy Phosphates on Beef Aging and Tenderness." *Journal of Food Science,* 38(1):56, 1973.

Kang, C.K., and W.D. Warner. "Tenderization of Meat with Papaya Latex Proteases." *J. Food Sci.,* 39:812, 1974.

Karmas, Endel. *Processed Meat Technology*. Park Ridge, New Jersey: Noyes Data Corp., 1976.

Kelly, Hugh J. *Food Service Purchasing—Principles & Practices*. New York: Chain Store Publishing Corp., 1976.

Komarik, S.L., D.K. Tressler, and L. Long. *Food Products Formulary*, Vol. 1 *Meat, Poultry, Fish and Shellfish*. Westport, Connecticut: Avi Publishing Co., 1974.

Korshagen, Bernice M., Ruth E. Baldwin, and Sue Snider. "Quality Factors in Beef, Pork and Lamb Cooked by Microwaves." *J. American Dietetic Association*, 69:635–639, December, 1976.

Kotschevar, Lendal H. *Quantity Food Purchasing*, 2nd ed. New York: Wiley, 1975.

Kramer, A., and B.A. Twigg. *Quality Control for the Food Industry*, 3rd ed. Vol. 1, *Fundamentals*. Vol. 2, *Applications*. Westport, Connecticut: Avi Publishing Co., 1970, 1973.

Laakkonen, et al. "Collagenolytic Activity and Water Holding Capacity as Factors Affecting Tenderness Using Low Temperature, Long Time Heating of Bovine Muscles." Doctoral Thesis for Cornell University, 1969.

Laakkonen, E., G.H. Wellington, and J.W. Sherbon. "Low-Temperature, Long-Time Heating of Bovine Muscle. 1. Changes in Tenderness, Water-Binding Capacity, pH, and Amount of Water-Soluble Components." *J. Food Sci.*, 35:175, 1970.

Lanier, T.C., J.A. Carpenter, and R.T. Toledo. "Effects of Cold Storage Environment on Color of Exposed Lean Beef Surfaces." *J. Food Sci.*, 42:860–864, 1977.

Lawrie, R.A. *Meat Science*, 3rd ed. Elmsford, N.Y.: Pergamon Press, 1979.

Levie, Albert. *Meat Handbook*, 4th ed. Westport, Connecticut: Avi Publishing Co., 1979.

Levinson, Charles. *Food and Beverage Operation—Cost Control and Systems Management*. Englewood Cliffs, New Jersey: Prentice-Hall, 1976.

Locker, R.H., C.L. Davey, P.M. Nottingham, D.P. Haughey, and N.H. Law. "New Concepts in Meat Processing." *Adv. Food Res.*, 21:158, 1975.

Locker, R.H., and C.J. Hagyard. "A Cold Shortening Effect in Beef Muscles." *J. Sci. Food Agric.*, 14:787, 1963.

Lorenz, K. "Microwave Heating of Foods—Changes in Nutrient and Chemical Compositions." *CRC Rev. Food Sci. Nutr.*, 7(4):339, 1976.

Manohar, S.V., and D.L. Rigby. "Effect of Sodium Tripolyphosphates on Thaw Drip and Taste of Fillets of Some Freshwater Fish." *Journal of Fisheries Research Board of Canada*, Vol. 30, pp. 685–688.

Marsh, B.B. "The Nature of Tenderness." *Proceedings 30th Annual Reciprocal Meat Conference of the American Meat Science Association*. Chicago: National Live Stock and Meat Board, 1977.

Marsh, B.B. "Temperature and Postmortem Change: Energy Use and Meat Quality." *Proceedings of the Meat Industry Research Conference*, March, 1977.

Miller, S.G. "Mechanical Tenderization of Meat in the HRI Trade." *Proceedings 28th Annual Reciprocal Meat Conference of the American Meat Science Association*. Chicago: National Live Stock and Meat Board, 1975.

Moeller, Fields, Dutson, Landmann, and Carpenter. "High Temperature Effects on Lysosomal Enzyme Distribution and Fragmentation of Bovine Muscle." *J. Food Sci.*, 42(2):510, 1977.

Mountney, G.J., *Poultry Products Technology*, 2nd ed. Westport, Connecticut: Avi Publishing Co., 1976.

NOAA/NMFS. *Current Fisheries Statistics*. U.S. Government Printing Office, Washington, D.C.

NOAA/NMFS. *Shellfish Situation and Outlook and Shellfish Market Review and Outlook*. Current Economic Analysis Series, U.S. Government Printing Office, Washington, D.C.

National Association of Meat Purveyors. *Meat Buyer's Guide*.

National Live Stock and Meat Board. *Meat in the Foodservice Industry*. Chicago, 1975.

National Live Stock and Meat Board. *Variety Meats for the U.S.A.—A Purchasing Guide*. Chicago, 1979.

Paul, P., and A.M. Child. "Effect of Freezing and Thawing Beef Muscles Upon Press Fluid, Losses and Tenderness." *Food Res.*, 2:339, 1937.

Pearson, G.L., and C. Bodine, Jr. "Boxed Beef Dispute: Cost Comparisons of Box vs. Carcass." *Meat Industry*, September, 1977.

Pierson, M.D., D.L. Collins-Thompson, and Z.J. Ordal. "Microbiological, Sensory and Pigment Changes of Aerobically and Anaerobically Packaged Beef." *Food Technology*, 24:129–133, 1970.

Potter, Norman N. *Food Science*, 2nd ed. Westport, Connecticut: Avi Publishing Co., 1973.

Price, J.F., and B.S. Schweigert. *The Science of Meat and Meat Products*, 2nd ed. San Francisco: W.H. Freeman, 1960.

Sacharow, S. *Handbook of Package Materials*. Westport, Connecticut: Avi Publishing Co., 1976.

Sale, A.J.H. "A Review of Microwaves for Food Processing." *J. Food Technology*, November, 1975, pp. 319–329.

Seideman, S.C., Z.L. Carpenter, G.C. Smith, and K.E. Hoke. "Effect of Degree of Vacuum and Length of Storage on the Physical Characteristics of Vacuum Packaged Beef Wholesale Cuts." *J. Food Sci.*, 41:732, 1976.

Seideman, S.C., Z.L. Carpenter, G.C. Smith, C. Vanderzant, and K.E. Hoke. "A Comparison of Vacuum Packaging Systems and Films on the Physical Characteristics of Beef Cuts." *J. Milk Food Technology*, 39:740–744, 1976.

Solomon, Kenneth I., and Norman Katz. *Profitable Restaurant Management*. Englewood Cliffs, New Jersey: Prentice-Hall, 1974.

Smith, G.C. "Factors Affecting the Palatability of Beef." *Proceedings of the Great Plains Livestock Conference*, 1975. Hastings, Nebraska, pp. 17–21.

Smith, G.C., T.C. Arango, and Z.L. Carpenter. "Effects of Physical and Mechanical Treatments on the Tenderness of the Beef Longissimus." *J. Food Sci.*, 36:445, 1971.

Smith, G.C., Z.L. Carpenter, and G.T. King. "Considerations for Beef Tenderness Evaluation." *J. Food Sci.*, 34:612, 1969.

Stadelman, William J. "A Method to Measure and Predict Meat Tenderness." *J. Anim. Sci.,* 32(5):80, 1978.

Stokes, John W. *How to Manage a Restaurant or Institutional Food Service,* 3rd ed. 1979.

Villano, Caesar, C.P.A., with Gloria B. Eagen. *Food Service Management and Control—The Profitable Approach.* New York: Lebhar-Friedman Books, 1977.

Warfel, M.C., and Frank H. Waskey. *The Professional Food Buyer—Standards, Principles, and Procedures.* Berkeley, California: McCutchan Pub. Corp., 1979.

Young, William G. "Vacuum Packaging Trends." *Modern Packaging*, April, 1975, pp. 185–186.

INDEX